Introduction to Chemical Principles

A Laboratory Approach

Seventh Edition

Susan A. Weiner
Department of Chemistry
West Valley College
Saratoga, California

Blaine Harrison
Department of Chemistry
West Valley College
Saratoga, California

BROOKS/COLE
CENGAGE Learning

Australia • Brazil • Japan • Korea • Mexico • Singapore • Spain • United Kingdom • United States

BROOKS/COLE
CENGAGE Learning™

**Introduction to Chemical Principles:
A Laboratory Approach, Seventh Edition**
Susan A. Weiner and Blaine Harrison

Publisher: Mary Finch

Acquisitions Editor: Lisa Lockwood

Development Editor: Stefanie Beeck

Assistant Editor: Stefanie Beeck

Editorial Assistant: Elizabeth Woods

Media Editor: Lisa Weber

Marketing Manager: Nicole Hamm

Marketing Assistant: Kevin Carroll

Marketing Communications Manager: Linda Yip

Project Manager, Editorial Production: Michelle Cole

Creative Director: Rob Hugel

Art Director: John Walker

Print Buyer: Judy Inouye

Permissions Editor: Bob Kauser

Production Service: Macmillan Publishing Solutions

Copy Editor: Aninda Bose

Cover Designer: Denise Davidson

Cover Image: Burning Methane Hydrate
© J. Pinkston and L. Stern / U. S. Geological Survey

Compositor: Macmillan Publishing Solutions

For product information and technology assistance, contact us at
Cengage Learning Customer & Sales Support, 1-800-354-9706.

For permission to use material from this text or product, submit all requests online at **cengage.com/permissions.**
Further permissions questions can be e-mailed to
permissionrequest@cengage.com.

Library of Congress Control Number: 2008942438

ISBN-13: 978-0-495-11479-6

ISBN-10: 0-495-11479-0

Brooks/Cole
10 Davis Drive
Belmont, CA 94002-3098
USA

Cengage Learning is a leading provider of customized learning solutions with office locations around the globe, including Singapore, the United Kingdom, Australia, Mexico, Brazil, and Japan. Locate your local office at
international.cengage.com/region.

Cengage Learning products are represented in Canada by Nelson Education, Ltd.

For your course and learning solutions, visit **academic.cengage.com.**

Purchase any of our products at your local college store or at our preferred online store **www.ichapters.com.**

Printed in the United States of America
1 2 3 4 5 6 7 12 11 10 09

Contents

Ⓜ *indicates microscale (small scale) experiments*

Preface

This laboratory manual is designed for several different types of chemistry courses. It can be used for a one-semester or one-quarter introductory chemistry course, a general chemistry course for nonscience majors, or a two-semester chemistry course designed for health science majors. There are a wide variety and a great number of experiments from which an instructor can assemble a customized course. As with earlier editions, this manual is suitable for use with several different textbooks.

A major change to this edition is an increase in the number of experiments pertaining to organic chemistry and biochemistry. Experiments 2 and 17 from the sixth edition have been omitted.

Several experiments are microscale (small scale) experiments. These are marked with the symbol Ⓜ in the Table of Contents for easy identification. Using small quantities of reagents increases safety, reduces the cost and disposal of chemicals, and allows for shorter completion times.

As in previous editions, the data and report sheets are printed in duplicate, one identified as Work Page and the other as Report Sheet. The students are directed to enter data into the Work Page during the experiment, then copy the finished data and calculations into the Report Sheet. This results in a clean and neat report.

Report sheets are designed to give adequate presentation of observations and results, but short enough to be graded easily and rapidly. Each experiment is independent, except for Experiments 20 and 21. All experiments can be completed in a three-hour laboratory period, including a prelaboratory discussion.

The Instructor's Manual includes lists of chemicals and equipment, data to be expected, answers to Advance Study Assignments, and miscellaneous suggestions.

A student finishing a laboratory program based on this manual will have become familiar with many laboratory operations and will have learned to collect and analyze experimental data. These skills will be a strong foundation for further work in general chemistry or other college-level science curriculum.

Susan A. Weiner
Blaine Harrison

Safety in the Laboratory

A chemistry laboratory can be, and should be, a safe place to work. Accidents can be prevented if you think about what you are doing at all times, use good judgment, observe safety rules, and follow directions. In addition to the rules below, comments appear in each experiment to alert you to probable hazards, including specific instructions on how to protect yourself and others against injury. Be sure to read these and keep the warnings in mind as you perform each experiment. Do not deviate from the procedures given in this book unless you are instructed to do so.

THERE IS NO SUBSTITUTE FOR SAFETY IN THE LABORATORY. Learn and observe these safety rules at all times:

1. Eye protection (OSHA approved goggles, safety glasses) must be worn by all students when working in the laboratory. This includes cleanup times and times when you yourself may not be working on an experiment, but someone near you is.

2. Do not eat or drink in the laboratory.

3. Do not taste any chemical.

4. Purses, sweaters, lunch bags, backpacks, and extra books should be stored in designated areas, but not in the laboratory working area. Backpacks, in particular, should not be on the floor near your laboratory desk.

5. Shoes must be worn in the laboratory at all times. Bare feet are prohibited.

6. Long hair should be tied back or pinned up, so it will not fall into chemicals or flames.

7. Do not work in the laboratory alone. An instructor or teaching assistant must be present.

8. Never perform any unauthorized experiment.

9. If an accident occurs in the laboratory, no matter how minor, report it to the instructor immediately.

10. All experiments or operations producing or using chemicals that release poisonous, harmful, or objectionable fumes or vapors MUST be performed in the fume hood.

11. Never point the open end of a test tube at yourself or at another person.

12. If you want to smell a substance, do not hold it directly to your nose; instead, hold the container a few centimeters away and use your hand to fan the vapors toward you.

13. Hot glassware and cold glassware look alike. If you heat glass and put it down to cool, do not pick it up too soon. Do not put hot glassware where another person is apt to pick it up.

14. When inserting a glass tube, rod, or thermometer into a rubber tube or stopper, protect your hands by holding the material with gloves or layers of paper towel. Lubricating the glass with water or glycerine is helpful.

15. When diluting acids, always add the acid to water, never water to the acid.

16. Most organic solvents are flammable. Keep these liquids away from open flames.

17. Do not pour organic solvents down a sink in the open laboratory. Dispose of them as directed by your instructor, or down a drain in the fume hood. Flush with plenty of water.

18. When disposing of liquid chemicals or solutions in the sink, flush with large quantities of water.

19. Do not wind the electric cord around a hot plate if it is still warm. The hot plate might melt the rubber insulation.

20. Do not dispose of matches, paper, or solid chemicals in the sink. Matches, after you are sure they are extinguished, and paper should be discarded into a wastebasket. Solid chemicals should be disposed of in whatever facility is provided in your laboratory.

21. Do not put broken glassware into wastebaskets. Dispose of it in designated places.

22. If you should have skin contact with any harmful chemical, flush the contact area with large quantities of water. Have a nearby student call the instructor for aid.

23. If you spill any chemical, solid or liquid, be sure to clean it up so another student does not come into contact with it and perhaps be injured by it.

24. Chemical characteristics, hazard levels, and safety instructions for the chemicals you use in the laboratory are described in Material Safety Data Sheets (MSDS) that are generally available in the laboratory. Follow directions given by your instructor in regard to these sheets. Pay close attention to particular safety precautions your instructor talks about before you begin each experiment.

25. Before leaving the laboratory, wipe the desk top and wash your hands with soap and water.

PREVENTING CONTAMINATION OF CHEMICALS

To conduct experiments successfully, you must avoid contaminating the chemical reagents you use, or reagents that will be used by other students after you. The following procedures will help minimize the possibility of contamination:

1. After washing glassware, always use a final rinse of deionized or distilled water.

2. Avoid handling more than one reagent bottle at a time, so you do not interchange their stoppers by mistake.

3. When selecting a reagent bottle, read the label twice to be sure you have the chemical you want.

4. Do not lay tops of reagent bottles or stoppers on the laboratory bench.

5. Use separate spatulas to remove different solid chemicals from their bottles.

6. Never remove a liquid reagent from a stock bottle with an eye dropper. Pour a small portion into a clean, dry beaker, and use your eye dropper to remove the liquid from the beaker.

7. When a quantity of a chemical is removed from its original container, whether it is a solid or a liquid, *do not* return any excess to the stock bottle. Dispose of the unused portion as directed by your instructor.

8. Never weigh a chemical directly on a balance pan. Use a preweighed container. Weighing paper is acceptable for most solid chemicals.

9. Some chemicals react with some stoppers. If you are going to store a chemical or solution in a bottle other than its original container, be sure the stopper you select (glass, rubber, cork) is suitable for that substance.

10. Never leave a stock bottle uncovered. Be sure you cover the bottle with the proper cover.

Common Laboratory Equipment

Beaker

Erlenmeyer flask

Suction flask

Graduated cylinder

Thermometer

Test tube

Buret

Pipet

Test-tube brush

Funnel

Büchner funnel

Crucible and cover

Eye dropper

Spot plate

Glass rod with rubber policeman

Common Laboratory Equipment

Crucible tongs

Test-tube holder

Bunsen burner
(Tirrill type)

Ring support

Utility clamp

Clay triangle

Ring stand with support

Buret clamp

Wire gauze

Evaporating dish

Watch glass

Tripod

Laboratory Procedures

The techniques found in many laboratory operations are so common that your instructions say simply, "Do this ..." with the assumption that you know exactly how to do it. For beginning students this assumption is often wrong. This may be your first opportunity to conduct a routine operation, and you may have questions about how to do it. This section discusses some of these methods.

HANDLING SOLID CHEMICALS

Your first step in taking a solid chemical is to read the label very carefully to be sure that you get the chemical you want. The names and formulas of different chemicals may be almost identical. For example, sodium sulfate is Na_2SO_4, and sodium sulfite is Na_2SO_3. The names differ by one letter, and the formulas differ by 1 in a subscript. We strongly recommend that you read all chemical names and formulas twice in the laboratory manual and twice again on the supply bottle.

When you need a chemical, take the container in which you will place it to the station from which the chemical is distributed. Transfer the chemical to the container there. Do not take a supply bottle to your work area.

Solid chemicals are generally distributed in wide-mouth, screw-cap bottles. If the substance is "caked" and doesn't flow easily, screw the cap on tightly and strike the bottle sharply against the palm of your hand. If this doesn't loosen the chemical, remove the cap and scrape the packed solid with the scoop you will use to remove the substance from the bottle. Having loosened the solid somewhat, you can often get it to flow freely by recapping the bottle and hitting it against your hand.

When you remove a cap from a bottle, place it on the desk with the top, or outside, of the cap down. This prevents contaminating the inside of the bottle from a dirty desk when the cap is returned to the bottle. Using a clean scoop, remove the amount of chemical you need. If you are transferring a closely controlled quantity of chemical, you can regulate the flow of solid from your scoop by holding the scoop over the receiving container and tapping your hand gently, as in Figure LP.1. If you have removed too much chemical, *do not return the excess to the bottle; instead throw it away*. The waste from this procedure is less of a problem than the contamination that will eventually occur if excess chemicals are returned to supply bottles. It follows that you should judge your requirement carefully and take no more chemical than you need.

After you have transferred the chemical you need, return the cap to the bottle and tighten it securely. If any solid has been spilled on the desk, clean it up before leaving the distribution station.

Figure LP.1
Transferring solid chemicals

HANDLING LIQUID CHEMICALS

The general procedures for handling solid chemicals apply to liquids, too. Specifically, (1) double-check the name and/or formula of the chemical you require and the chemical you get; (2) take your container to the distribution station, rather than taking the supply bottle to your work area; (3) do not place the cap or stopper of a supply bottle on the desk in such a way that the inside of the cap touches the desk; (4) if you remove too much liquid from the supply bottle, *do not return it, but throw it away;* (5) be sure to return the cap or stopper to the supply bottle when you are finished; and (6) wipe up any liquid that may have spilled.

Figure LP.2 shows the technique for controlling the flow of liquid from a bottle by pouring down a stirring rod. This technique may also be used when pouring from a beaker, as shown in Figure LP.3.

Liquids are frequently distributed from bottles fitted with eye droppers. When using a dropper for removing the liquid, be sure to hold the dropper vertically *with the rubber bulb at the top* so that the liquid does not

Figure LP.2
Pouring liquids from a bottle

Figure LP.3
Pouring liquids from a beaker

drain into the bulb and become contaminated. If you are required to take a small quantity of liquid from a bottle not fitted with a dropper and wish to use your own, do not place your dropper into the supply bottle. The proper procedure is to pour some of the liquid into a small beaker and then use your dropper to transfer the liquid from the beaker to your container. Excess liquid should be thrown away, as noted above. Estimate your needs carefully so the excess can be kept to a minimum.

Many liquids used in the laboratory are flammable, many release harmful vapors, and many have both of these dangerous properties. When working with such chemicals, it is best to work in a fume hood. When disposing of such chemicals, always follow the specific procedures established in your laboratory. If a liquid is flammable, do not use it anywhere near an open flame. Vapors from your liquid could drift to the flame and become an invisible wick by which the flame could travel right back to your liquid and cause a fire.

QUANTITIES OF CHEMICALS

Most chemical quantities identified in this book are approximate quantities that are practical for the sizes of beakers, test tubes, and other containers you will use. If the quantity you take falls within 10% of the amount called for, it will be satisfactory. It is therefore unnecessary for you to try to measure out "exactly" the amount specified. In fact, trying to get that exact amount is a waste of time, both your time and the time of other students who will be delayed because you tie up a balance.

While using an exact quantity of a chemical is not important, knowing as accurately as possible the quantity actually used is essential if that quantity becomes a part of your calculations. You will recognize this requirement if your instructions call for so many milliliters of a liquid "estimated to the nearest 0.1 mL," or to "measure 1.5 grams of a solid on a milligram balance." The first tells you to pour into a graduated cylinder a quantity of liquid that is within about 10% of the amount specified, and then to measure and record that quantity to the nearest 0.1 mL. The second instruction may be interpreted as, "Take between 1.35 and 1.65 grams of a chemical and then measure and record the quantity taken to the nearest milligram."

Several experiments in this book require "about 1 to 2 mL" of a liquid, usually to be placed in a test tube. Again the exact quantity is not important, and it is a waste of time to measure it with a graduated cylinder. Most eye droppers deliver drops of such size that there are about 20 drops to the milliliter; and the total volume drawn into a dropper by one squeeze of the bulb is about 1/2 milliliter. One milliliter therefore can be estimated simply as two droppers-full.

READING VOLUMETRIC GLASSWARE

When a liquid is placed into a glass container it forms a **meniscus,** a curved surface that is lower in the middle than at the edge. Volumetric laboratory equipment is calibrated to measure volume by sighting to the *bottom* of the meniscus, as shown in Figure LP.4. Notice that it is essential that the line of sight be perpendicular to the calibrated vessel if you are to read it accurately. It is also important that you hold the vessel vertically.

Four types of calibrated glassware are used in the experiments in this book. The most accurately calibrated are the volumetric pipets and burets used in Experiments 20 and 21. Most of your volume measurements will be made in graduated cylinders. Their main purpose is to measure volumes and they are designed and calibrated accordingly. Beakers and Erlenmeyer flasks made by some manufacturers are also "calibrated," even though the function of these items has nothing to do with measuring volume. The calibrations on beakers and flasks give only *very rough* indications of volume up to a certain level in the vessel. Volumes estimated by these calibrations should *never* be used in calculations.

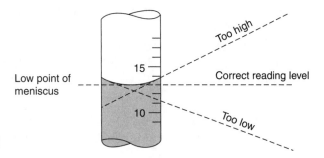

Figure LP.4
Reading the volume of a liquid

MEASURING MASS—WEIGHING

You will no doubt receive specific instructions on the use of chemical balances in your laboratory. No attempt will be made to duplicate those instructions here. Instead, comments will be limited to some general suggestions, plus identification of a term that has special meaning throughout this book.

Chemicals are *never* weighed directly on the pan of a laboratory balance. Instead, the mass is determined by a process known as **weighing by difference.** A suitable container—a small beaker, or perhaps a test tube that is to be used in the experiment—is weighed empty on the balance. The desired chemical is added to the container, and the total mass of the combination is determined. By subtracting the mass of the empty container from the mass of the container plus chemical, you find the mass of the chemical.

Throughout this book the word *container* is used to include any and all objects that pass through the entire experiment unchanged in mass. In addition to a test tube, for example, you might include in the mass of the "container" a test-tube holder by which the test tube is suspended on a balance during weighing, or the mass of a beaker in which the test tube is held for weighing. In one experiment the mass of a liquid is measured in a graduated cylinder that is covered with a piece of plastic film. The film is weighed with the empty cylinder, and their combined masses make up the mass of the "container." In the various experiments where you see the "container" identified, the word has the meaning given in this paragraph.

Sometimes students use containers that are not actually part of the experiment in taking samples of solid chemicals. Most common is the practice of placing a piece of paper on the pan of a balance, transferring the required quantity of chemical to the paper, and then transferring it to the vessel to be used in the experiment. If you use this technique to obtain a measured mass of the chemical, your first weighing should be of the paper with the chemical on it. Then transfer the chemical, and bring the paper back for a second weighing. This way your difference will be the mass of the chemical actually transferred, unaffected by any chemical that may have remained on the paper unnoticed. In this method you should use a hard, smooth paper—waxed paper is best—rather than coarse paper, such as paper towel, which is certain to trap powders and tiny crystals.

Laboratory balances are subject to corrosion. Both the balances and the balance area should be kept clean, and spilled chemicals should be cleaned up immediately.

Here are a few miscellaneous pointers on proper balance operation, given as a series of "do's and don'ts," with some items in both lists for emphasis:

DO   Allow hot objects to cool to room temperature before weighing.

Close the side doors or hood of a milligram balance while weighing.

Record all digits allowed by the accuracy of the balance used, even if the last digit happens to be a zero on the right side of the decimal point.

DON'T Weigh objects that are warm or hot.

Weigh objects that are wet (evaporation of water will change the mass).

Weigh volatile liquids in uncovered vessels.

Touch the object with your hand if you are using a milligram or analytical balance; your fingerprints have weight, too!

Forget to check the zero on a milligram balance before and after weighing.

Forget to record the mass to as many digits as the accuracy of the balance allows—and no more.

LABORATORY BURNERS

The function of a laboratory burner is to provide an adjustable mixture of natural gas and oxygen (from the air) that may be burned to produce the kind of flame required for a specific purpose. As a group the burners are called Bunsen burners, although most burners used today are improvements over the original Bunsen design. All have the same general features, and the Tirrill burner described in Figure LP.5 is representative of the group.

Gas enters the barrel of the burner from the center of the base, controlled by a valve in the base. Air enters through an opening at the bottom of the barrel where it screws onto the base. The amount of air admitted is

Outer cone

Region of highest temperature

Bright blue cone (combustion zone)

Inner dark cone

Barrel

Air vents

Gas inlet

Needle valve (for gas control)

Figure LP.5
Laboratory burner (Tirrill type) and flame

governed by the position of the barrel. When the barrel is screwed down, the air opening is small. This limits the amount of air, and for a given amount of gas it gives a mixture that has a high gas-to-air ratio—a "rich" mixture. If the barrel is unscrewed to admit a large quantity of air, the mixture has a low gas-to-air ratio—a "lean" mixture. By adjusting the amount of gas at the bottom, you control the size of the flame; and by adjusting the amount of air with the barrel, you control the type of flame produced.

If you burn a mixture with very little air—a very high gas-to-air ratio, or a very rich mixture—the flame will be yellow and not very hot. The yellow color is from unburned carbon, which is deposited as soot on the bottom of any vessel that is heated with such a flame. Increasing the amount of air causes the flame to become less yellow and more blue, and finally all blue. As still more air is introduced, the blue flame separates into parts, a light-blue inner cone and a darker outer cone. The hottest part of the flame is just above the tip of the bright blue cone. If too much air is introduced, the entire flame will "rise" and burn noisily above the burner barrel.

Occasionally, a burner will "strike back" and burn the mixture inside the barrel where the two components first meet. You usually become aware of this condition by the noise produced in burning. If this happens, shut the burner off briefly, and then relight it. Be careful, however, because the barrel of a burner that is striking back becomes *very hot*.

If you are not familiar with laboratory burners, it is recommended that you light one and experiment with the various adjustments to see how they work. Don't be afraid of a burner. It is a simple device that cannot hurt you unless you put your hand in the flame or touch the barrel of a burner that has been striking back. It is also a rugged device that you will not damage without trying to.

The proper lighting procedure is to strike the match, open the gas valve at the laboratory desk completely, and then move the match flame to the burner just below the tip of the barrel, letting the top of the flame creep over the top of the barrel to light the gas. If your burner has no gas control valve in its base, you will have to control the gas at the desk; otherwise the desk valve is opened fully, and the gas flow is adjusted at the burner.

When you first use a burner to heat a cold object, start with a blue flame, but one that is not very strong, or the object may crack. A blue flame that has no inner cone is ideal. After about a minute you can increase the amount of air in the flame to produce an inner cone and higher temperature. Crucibles and, with special precautions, test tubes may be heated directly in the flame. Crucibles are usually mounted on a clay triangle directly over the flame, and test tubes are held by hand in test-tube clamps. When you are heating a liquid in a beaker or flask, the vessel should be placed on a wire screen with an asbestos center, which is mounted on a ring stand or tripod. It is permissible to heat beakers and flasks made of Pyrex or other heat-resistant glass; but never, *under any circumstances*, heat calibrated volumetric glassware, such as graduated cylinders. There are two reasons for this "don't." First, the glass is not heat resistant and will probably crack. Second, heating would expand the glass and probably destroy the accuracy of the calibration.

PLACING RUBBER STOPPERS ON GLASS TUBING OR ROD

This book contains no experiment in which it is necessary to place a glass tube or thermometer through a rubber stopper to produce an airtight fit. We will therefore not describe the precautions that must be observed in this procedure, but rather describe the alternative procedure that may be used if the fit does not have to be airtight. A rubber stopper with one or more holes is cut from the side to one of the holes. The split stopper may then be held open, as shown in Figure LP.6, and the tubing or thermometer slipped into place. When the stopper is released it closes around the glass, holding it firmly in place.

Figure LP.6
The use of a split stopper

Properties and Changes of Matter

Performance Goals

1–1 Determine experimentally the solubility of a pure substance in a given liquid, or, in the case of two liquids, determine their miscibility.

1–2 Determine experimentally which of two immiscible liquids is more dense.

1–3 Determine whether or not a chemical reaction occurs when you combine two solutions, and state the evidence for your decision.

CHEMICAL OVERVIEW

All material things that compose our universe are referred to as **matter.** Matter is commonly defined as that which has mass and occupies space. In this experiment you will examine some of the characteristics of matter and be introduced to some of the language of science in which these characteristics are described.

A **pure substance** is a sample of matter that has identical properties throughout, and a definite, fixed composition. **Physical properties** are those characteristics of a substance that can be observed without changing the composition of the substance. Common physical properties are taste, color, odor, melting and boiling points, solubility, and density. **Chemical properties** describe the behavior of a substance when it changes its composition by reacting with other substances or decomposing into two or more other pure substances. The ability to burn and the ability to react with water are chemical properties.

Matter can undergo two types of changes, physical and chemical. **Physical changes** do not cause a change in composition, only in appearance. For example, when copper is melted, only a change of state occurs; no new substance is formed. In a **chemical change,** substances are converted into new products having properties and compositions that are entirely different from those of the starting materials. Wood, for example, undergoes a chemical change when it burns by reacting with oxygen in the air, forming carbon dioxide and water vapor as the new products.

When two liquids are mixed, the mixture may be completely uniform in appearance. In this case the liquids are said to be **miscible.** Some liquids are miscible in all proportions, while others have a limited range of miscibility. If the two liquids are not at all miscible, i.e., **immiscible,** two distinct layers will form when they are poured together. The liquid having the lower density will "float" on top of the other.

When a solid is added to and dissolves in a liquid, it is **soluble** in that liquid. The mixture formed is called a **solution.** A liquid solution is always clear; it may be colorless, or it may have a characteristic color. If the solid does not dissolve, it is said to be **insoluble.**

When two solutions are combined, a chemical change, or **reaction,** may occur in which new products form. If so, it will be evidenced by one of several visible changes. Among them are:

1. *Formation of a* **precipitate**, *or a solid product.* A precipitate is often very finely divided and distributed throughout the solution, giving a "cloudy" appearance. If allowed to stand, the precipitate will settle to the bottom of its container. The precipitate may be separated from the liquid by passing the mixture through a *filter* that collects the solid particles, but permits the solution to pass through.

2. *Formation of a gaseous product.* The gas produced bubbles out of the solution, a process called **effervescence.**

3. *Occurrence of a color change.* Usually a color change indicates the formation of a product with a color not originally present among the reactants. Sometimes the color will be the same as that of one of the reactants, but a darker or lighter shade.

In many cases no reaction occurs when two solutions are brought together.

SAFETY PRECAUTIONS AND DISPOSAL METHODS

Fumes from trichloroethane, xylene, and ammonia solutions are potentially harmful. Confine your use of these liquids to the fume hoods. Skin contact with these three liquids, or with hydrochloric acid, should be avoided. If it occurs, rinse the affected area thoroughly with water, and then wash with soap and water. Be sure to wear approved eye protection throughout the experiment.

Trichloroethane and xylene mixtures should be collected in stoppered bottles. **Do not pour them down the drain.** Solutions containing heavy metal precipitates should be collected in a separate container.

PROCEDURE

1. Mixing Liquids

A. Place about 20 drops of trichloroethane into a small-size test tube. Note this approximate quantity, because you will have several occasions in the experiment to estimate this volume in a test tube. Add about 10 drops of water and gently shake the test tube, or mix the contents with a stirring rod. Are the two liquids miscible? Record your observation on the work page.

If the two liquids are *not* miscible, identify the liquid, trichloroethane or water, that is on top. You may determine which liquid is on top by the relative quantities placed into the test tube: you added two times as much trichloroethane as you did water. Record the name of the top liquid on your work page.

B. Discard the mixture from Part 1A and repeat the experiment, this time using about 20 drops of water first, followed by 10 drops of trichloroethane. If the liquids are not miscible, again record on the work page which liquid is on top.

C. Using a clean test tube, or the original one thoroughly rinsed with water, repeat the procedure with about 20 drops each of methanol (methyl alcohol) and water. It is not necessary this time or hereafter to reverse the order of liquids, as in Parts 1A and 1B. Again record your observations and conclusions.

D. Using a clean and thoroughly rinsed test tube, repeat the procedure again, this time with about 20 drops each of water and xylene. Record your observations and conclusions as before.

E. Using a clean *and dry* test tube (there must be no water present), perform the experiment once again, now using about 20 drops each of trichloroethane and xylene. Record your observations.

2. Dissolving a Solid in a Liquid

In this part of the experiment and the next, you will be preparing solutions. The procedure is to take about 4 mm—just over 1/8 inch—of the solid on the tip of a spatula and place it into about 10 mL of deionized (or distilled) water in a test tube. Shake the test tube gently, or stir the contents with a clean, dry stirring rod. If none of the solid appears to dissolve, the substance is insoluble. If *any* of it dissolves, but a small amount does not, add more water to get all of the solid into the solution.

A. Place a small quantity of barium chloride, $BaCl_2$, in water as described above. Does the solid dissolve in the water? Record your observations and save the solution for further use.

B. Add a small amount of sodium sulfate, Na_2SO_4, to about 10 mL of water in a second test tube. Does the solid dissolve? Record your observations and save the solution.

C. Combine the contents of the test tubes from Steps 2A and 2B in a large test tube. Record your observations. Set the test tube aside for 5 to 10 minutes and examine it again. Record what you see.

D. Add a small amount of barium sulfate, $BaSO_4$, to about 10 mL of water. Is this compound soluble? Record your observations.

3. Mixing Solutions

A. In a small test tube, dissolve a small amount of iron(III) chloride, $FeCl_3$, in about 2 mL of water. In another test tube, dissolve a small amount of potassium thiocyanate, KSCN, in about 2 mL of water. Mix the two solutions and record your observations.

B. In a small test tube, dissolve a small amount of sodium chloride, NaCl, in about 2 mL of water. In another test tube, dissolve a small amount of ammonium nitrate, NH_4NO_3, in about 2 mL of water. Mix the two solutions and record your observations.

C. In a small test tube, dissolve a small amount of sodium carbonate, Na_2CO_3, in about 2 mL of water. Add 2 to 3 drops of hydrochloric acid, HCl, watching carefully for any evidence of a chemical reaction. Then add some more HCl and watch for a reaction. Record your observations.

D. In a small test tube, dissolve a small amount of calcium chloride, $CaCl_2$, in about 2 mL of water. In another test tube, dissolve a small amount of sodium carbonate, Na_2CO_3, in about 2 mL of water. Mix the two solutions and record your observations.

E. In a small test tube, dissolve a small amount of copper(II) sulfate, $CuSO_4$, in about 2 mL of water. Add concentrated ammonia solution, $NH_3(aq)$, to it, a drop at a time. (Ammonia solutions are sometimes labeled ammonium hydroxide, NH_4OH.) Record your observations.

Experiment 1

Advance Study Assignment

1. Distinguish between physical and chemical properties. Give an example of each.

2. Classify each of the following as a physical or chemical change:

   a. Iron rusting _____   **c.** Burning a candle _____

   b. Boiling water _____   **d.** Silver tarnishing (turning black) _____

3. Identify three forms of evidence that a chemical reaction has occurred:

   a.

   b.

   c.

4. Define a precipitate.

Experiment 1

Work Page

Part 1—Mixing Liquids

Step	Mixture	Miscible or Immiscible	Liquid on Top	More Dense Liquid
1A	Trichloroethane (20 drops) Water (10 drops)			
1B	Water (20 drops) Trichloroethane (10 drops)			
1C	Water Methanol			
1D	Water Xylene			
1E	Trichloroethane Xylene			

Questions

1. Is it possible from Parts 1A–1D to determine which of the liquids, xylene or trichloroethane, is more dense? If so, identify the liquid with the greater density and explain how you reached your conclusion; if not, explain why.

   More dense liquid: _____

   Explanation:

2. Is it possible from Part 1E alone to determine which of the liquids, xylene or trichloroethane, is more dense? If so, identify the liquid with the greater density and explain how you reached your conclusion; if not, explain why.

   More dense liquid: _____

   Explanation:

Part 2—Dissolving a Solid in a Liquid

2A. Barium chloride: Soluble (_____) or insoluble (_____)?

2B. Sodium sulfate: Soluble (_____) or insoluble (_____)?

2C. Mixture of contents of test tubes from 2A and 2B:
Immediate appearance:

Appearance 5–10 minutes later:

2D. Barium sulfate: Soluble (_____) or insoluble (_____)?

Optional. Based on your observations from Steps 2A, 2B, and 2D, suggest an explanation for your observation in Step 2C.

Part 3—Mixing Solutions

Step	Solutions Combined	Reaction: Yes or No	Evidence
3A	Iron(III) chloride Potassium thiocyanate		
3B	Sodium chloride Ammonium nitrate		
3C	Sodium carbonate Hydrochloric acid		
3D	Calcium chloride Sodium carbonate		
3E	Copper(II) sulfate Ammonia solution		

Experiment 1

Report Sheet

Part 1—Mixing Liquids

Step	Mixture	Miscible or Immiscible	Liquid on Top	More Dense Liquid
1A	Trichloroethane (20 drops) Water (10 drops)			
1B	Water (20 drops) Trichloroethane (10 drops)			
1C	Water Methanol			
1D	Water Xylene			
1E	Trichloroethane Xylene			

Questions

1. Is it possible from Parts 1A–1D to determine which of the liquids, xylene or trichloroethane, is more dense? If so, identify the liquid with the greater density and explain how you reached your conclusion; if not, explain why.

   More dense liquid: _____

   Explanation:

2. Is it possible from Part 1E alone to determine which of the liquids, xylene or trichloroethane, is more dense? If so, identify the liquid with the greater density and explain how you reached your conclusion; if not, explain why.

   More dense liquid: _____

   Explanation:

Part 2—Dissolving a Solid in a Liquid

2A. Barium chloride: Soluble (_____) or insoluble (_____)?

2B. Sodium sulfate: Soluble (_____) or insoluble (_____)?

2C. Mixture of contents of test tubes from 2A and 2B:
Immediate appearance:

Appearance 5–10 minutes later:

2D. Barium sulfate: Soluble (_____) or insoluble (_____)?

Optional.   Based on your observations from Steps 2A, 2B, and 2D, suggest an explanation for your observation in Step 2C.

Part 3—Mixing Solutions

Step	Solutions Combined	Reaction: Yes or No	Evidence
3A	Iron(III) chloride Potassium thiocyanate		
3B	Sodium chloride Ammonium nitrate		
3C	Sodium carbonate Hydrochloric acid		
3D	Calcium chloride Sodium carbonate		
3E	Copper(II) sulfate Ammonia solution		

The Chemistry of Some Household Products

Performance Goals

2–1 Perform tests to confirm the presence of known ions in certain solids.

2–2 Perform tests to confirm the presence of these ions in household products.

2–3 Analyze and identify an unknown solid.

CHEMICAL OVERVIEW

In everyday life we encounter a large number of chemicals. In fact, everything that surrounds us is a chemical. The air we breathe is a mixture of gaseous elements, oxygen, and nitrogen. The food we eat contains carbohydrates, proteins, large molecular mass vitamins, etc. You may not think about it, but you handle all kinds of chemicals every day.

Qualitative analysis is used to identify components of a solution or solid. A reagent that causes an easily recognized reaction with a particular ion present is added to a sample of the unknown. If the reaction occurs, the ion is present; if the reaction does not occur, the ion is absent.

In this experiment you will perform tests on known compounds that show the presence of certain ions. Once you have become familiar with these specific reactions, you will perform the same tests on some common household products. Finally, you will be given an unknown compound. By performing a series of tests, you will determine which of the following ions is present in your unknown: Cl^-, NH_4^+, SO_4^{2-}, HCO_3^-, PO_4^{3-}.

SAFETY PRECAUTIONS AND DISPOSAL METHODS

Acids and bases are corrosive and contact with your skin should be avoided. Any spilled acid or base should be washed off promptly. Be sure to wear goggles or safety glasses while performing this experiment.

Dispose of excess solids as directed by your instructor. Solutions containing precipitates should be collected in a waste container.

PROCEDURE

1. Table Salt, NaCl

The presence of chloride ions, Cl^-, can be detected by reacting the dissolved substance with silver nitrate, $AgNO_3$. A white precipitate of silver chloride, $AgCl$, will form. This reaction is typical of all solutions that contain the Cl^- ion.

Place a few crystals of table salt in a small test tube. Dissolve the solid in about 10 drops of deionized water. Add 2 drops of 1 M nitric acid, HNO_3. Then add 2 or 3 drops of 0.1 M silver nitrate, $AgNO_3$. Record your observations.

2. Garden Fertilizers

Some of the active ingredients of ordinary garden fertilizers are ammonium salts. These compounds are the source of nitrogen, an element essential for the growth of plants. When a strong base, such as sodium hydroxide, $NaOH$, is added to a compound containing the NH_4^+ ion, gaseous ammonia, NH_3, is liberated. This can be detected by a piece of red litmus paper that has been moistened with deionized water. The paper will turn blue if NH_3 is present.

A. Pour about 10 drops of 1 M ammonium chloride, NH_4Cl, into a small test tube. Add about 10 drops of 3 M sodium hydroxide, $NaOH$. Hold a piece of moist red litmus paper in the mouth of the test tube. Do not allow the paper to come into contact with the side of the test tube, since it may have $NaOH$ on it. Record your observations. If you notice no change, gently warm the test tube in a hot-water bath and check with litmus paper again.

B. Place a small amount of garden fertilizer into a test tube. Add about 10 drops of 3 M $NaOH$ to the solid. Hold a moist strip of red litmus paper in the mouth of the test tube. Record your observations.

3. Epsom Salt, $MgSO_4 \cdot 7H_2O$

Epsom salt can be purchased in any drugstore. It is commonly used to prepare soothing baths, and it is sometimes used as a purgative. If a solution of barium chloride, $BaCl_2$, is added to a solution of Epsom salt, a finely divided white precipitate of barium sulfate, $BaSO_4$, will form.

A. Pour about 10 drops of 1 M sodium sulfate, Na_2SO_4, into a small test tube. Add 3 drops of 1 M hydrochloric acid, HCl, and 2 or 3 drops of 1 M $BaCl_2$. Record your observations.

B. Place a small amount of Epsom salt into a test tube. Dissolve the solid in about 10 drops of deionized water. Add 3 drops of 1 M HCl and 2 or 3 drops of 1 M $BaCl_2$. Record your observations.

4. Baking Soda, $NaHCO_3$

Baking soda and baking powder both contain sodium hydrogen carbonate, $NaHCO_3$. When this compound reacts with an acid, gaseous carbon dioxide is produced. This makes bread or cakes "rise." Carbon dioxide is produced whenever compounds containing HCO_3^- or CO_3^{2-} ions react with acids. The formation of bubbles indicates that at least one of these ions is present in a sample.

A. Using a spot plate, dissolve a small amount of baking soda in about 10 drops of deionized water. Add about 6 to 8 drops of 1 M hydrochloric acid, HCl. Record your observations. Repeat the test by dissolving a small amount of baking soda directly in HCl. Again record your observations, noting any difference between the two reactions.

B. Take another small amount of solid baking soda and add 10 drops of commercial vinegar to it. Record your observations, noting any difference between this reaction and the ones in Step 4A.

5. Detergents

One of the common ingredients of laundry detergents and wall-washing compounds is sodium phosphate, Na_3PO_4. The PO_4^{3-} ion can be detected by adding ammonium molybdate, $(NH_4)_2MoO_4$, to a dissolved sample. A yellow, powdery precipitate will form. Sometimes gentle heating in a water bath is necessary to hasten the reaction.

A. Pour about 10 drops of 1 M sodium phosphate, Na_3PO_4, into a small test tube and add 3 M HNO_3 until the solution is acidic. (Test by dipping a stirring rod into the solution and touching the wet rod to a strip of blue litmus paper. The solution is acidic if the color changes to red.) Then add 6 to 8 drops of 0.5 M $(NH_4)_2MoO_4$ and place the test tube in a hot-water bath for a few minutes. Record your observations.

B. Repeat the procedure with a small amount of laundry detergent. Dissolve the solid in about 10 drops of deionized water, acidify it, and add the molybdate reagent. Record your observations. Heat the solution in a water bath and note any change, if there is one.

6. Identification of an Unknown

Obtain a solid unknown and record its number on your work page. Place small portions of the unknown in five separate test tubes. Keep enough unknown to make three additional tests, in case you wish to repeat one or more of the procedures. Dissolve each solid portion in about 10 drops of deionized water and perform the five tests on the separate portions. Your unknown will contain only one ion. Identify the ion and record it on your work page. Remember, you have not identified the cation, so you cannot record your result as "Epsom salt," you have only identified the SO_4^{2-} ion! Record it as such.

Experiment 2

Advance Study Assignment

1. What would happen if you used tap water instead of deionized water in this experiment?

2. You have a piece of chalk. How would you determine if it contains $CaCO_3$?

3. If you have a white powder that could be either Epsom salt or table salt, how could you decide which one it is?

4. Why is it dangerous to mix household chemicals indiscriminately?

Experiment 2

Work Page

Substance Tested	
Table salt	
NH_4Cl	
Fertilizer	
Na_2SO_4	
Epsom salt	
Baking soda solution + HCl	
Baking soda solid + HCl	
Baking soda solid + vinegar	
Na_3PO_4	
Detergent	

Unknown number _____ Ion present _____

Describe the experimental observation on which your result is based:

Experiment 2

Report Sheet

Substance Tested	
Table salt	
NH$_4$Cl	
Fertilizer	
Na$_2$SO$_4$	
Epsom salt	
Baking soda solution + HCl	
Baking soda solid + HCl	
Baking soda solid + vinegar	
Na$_3$PO$_4$	
Detergent	

Unknown number _____ Ion present _____

Describe the experimental observation on which your result is based:

Separation of Cations by Paper Chromatography

Performance Goals

3–1 Separate a mixture of cations by paper chromatography and calculate their R_F values.

3–2 Analyze an unknown mixture of cations by paper chromatography.

CHEMICAL OVERVIEW

Chromatography, which means "the graphing of colors," gets its name from the early experiments of Tswett, who, in 1906, succeeded in separating a mixture of colored pigments obtained from leaves. A solvent mixture, carrying the pigments, was allowed to pass through a glass column packed with chalk. At the end of the experiment, the pigments were separated in colored bands at various distances from the starting level. This method is now known as column chromatography.

Chromatography may now be applied to colorless compounds and to ions. Paper chromatography is a more recent and much faster separation technique than column chromatography. It may be used for the separation of substances by a solvent moving on sheets or strips of filter paper. The filter paper is referred to as the **stationary phase,** or **adsorbent.** The mixture of solvents used to carry the substances along the paper is called the **mobile phase,** or **solvent system.**

In practice, a sample of the solution containing the substances to be separated is dried on the paper. The end of the paper is dipped into the solvent system so that the sample to be analyzed is slightly above the liquid surface. As the solvent begins to soak the paper, rising by capillary action, it transports the sample mixture upward. Each component of the mixture being separated is held back by the stationary phase to a different extent. Also, each component has a different solubility in the mobile phase and therefore moves forward at a different speed. A combination of these effects causes each component of the mixture to progress at a different rate, resulting in separation.

In a given solvent system, using the same adsorbent at a fixed temperature, each substance can be characterized by a constant **retention factor,** R_F. By definition,

$$R_F = \frac{\text{Distance from origin to center of spot}}{\text{Distance from origin to solvent front}} \qquad (3.1)$$

where the **origin** is the point at which the sample was originally placed on the paper and the **solvent front** is the line representing the most advanced penetration of the paper by the solvent system. The R_F value is a characteristic property of a species, just as the melting point is a characteristic property of a compound.

In this experiment you will separate a mixture of iron(III), copper(II), and cobalt(II) ions, Fe^{3+}, Cu^{2+}, and Co^{2+}, respectively. Each ion forms a different colored complex when sprayed with a solution containing potassium hexacyano ferrate(II), $K_4[Fe(CN)_6]$.

SAFETY PRECAUTIONS AND DISPOSAL METHODS

Acetone is *extremely flammable*. Its vapors can ignite even when the liquid is a considerable distance from an open flame, so be sure no such flame is operating in the vicinity of your work area. Fumes of acetone and concentrated hydrochloric acid are objectionable and, to some degree, harmful. These chemicals should be used in the hood. Be sure to wear safety glasses.

After you have finished the experiment, dispose of the solvent mixture in a stoppered bottle.

PROCEDURE

1. Using a graduated cylinder and working in the hood, prepare the following solvent system: 19 mL acetone; 4 mL concentrated hydrochloric acid, HCl; 2 mL water or use 25 mL of a pre-prepared solvent mixture, if available. Pour the solvent mixture into an 800- or 1000-mL beaker and cover it tightly with a plastic film (e.g., Saran wrap). This allows the atmosphere within the beaker to become saturated with solvent vapor and helps to give a better chromatographic separation.

2. Obtain a piece of chromatography paper 24 to 25 cm long by 11 to 14 cm wide. Draw a pencil line about 1 cm from the long edge of the paper. (You must use an ordinary pencil for this line. Ink or colored pencil often contains substances that may be soluble in the solvent, producing chromatograms of their own.) This line will indicate the origin (see Figure 3.1). Also draw a line about 1 cm long, 6 cm above the center of the penciled line.

3. Using a different capillary tube for each solution (do not mix them!) transfer a drop of each solution listed below to the penciled line, as shown in Figure 3.1. Apply the spots evenly over the line, leaving a margin of about 3 cm from each short edge of the paper. Use a separate, clean capillary tube for each solution; or, if the solutions are to be

obtained from beakers in which a capillary tube is provided, *be sure to return the tube to its proper beaker.* With a pencil, identify each spot by writing on the paper directly beneath the spot. The solutions are:

A. Fe^{3+} solution

B. Cu^{2+} solution

C. Co^{2+} solution

D. Solution containing all three ions, Fe^{3+}, Cu^{2+}, and Co^{2+}

E. Any of the unknowns furnished (be sure to record its number)

F. Another unknown (again, record the number).

4. Dry the paper under a heat lamp or air blower.

5. Form the paper into a cylinder without overlapping the edges. Fasten the paper with staples, as shown in Figure 3.2.

6. Place the beaker in a position on your desk where it will remain undisturbed throughout this step. Taking care that the origin line remains above the solvent, carefully place the cylinder into the beaker,

Figure 3.1
Preparing chromatography paper

Figure 3.2
Development of chromatogram

as shown in Figure 3.2. Replace the plastic film and wait as the solvent moves up the paper. Do not move the beaker or the solvent front will be uneven.

7. *NOTE: In this and all remaining steps, when the paper is wet, be sure not to lay it down on any surface that is not clean.* When the solvent has risen above the short line drawn 6 cm above the origin in Step 2, remove the cylinder from the beaker and quickly mark the solvent front position with a pencil. Remove the staples and dry the paper under a heat lamp.

8. Spray the paper with a solution of potassium hexacyano ferrate(II), $K_4[Fe(CN)_6]$.[*]

The presence of Fe^{3+} is shown by the spot turning a dark steel blue color. Cu^{2+} turns a rust brown, and Co^{2+} turns grayish purple (rose).

RESULTS AND CALCULATIONS

Observe and record on the work page the colors of spots produced by the three ions in each of the chromatograms of solutions A through D.

Measure and record in millimeters the distance between the origin and the solvent front (X in Figure 3.3). Next, measure and record the distance between the origin and the center of each spot in the chromatograms for solutions A through D. Calculate the R_F value for each ion, using Equation 3.1. Record that value as a decimal fraction to two decimals (i.e., 0.82, 0.75).

From the spots above your solutions E and F, indicate by a check mark in the table the ions present in the unknowns. Be sure to list the identification numbers of the unknowns.

If the center of a spot is right on the solvent front, the R_F value equals 1.00.

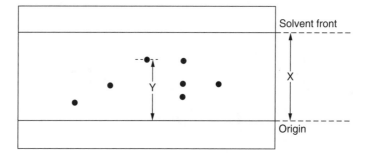

Figure 3.3
Developed chromatogram

---

[*]Also called potassium ferrocyanide.

Experiment 3

Advance Study Assignment

1. What would you observe if you used a ballpoint pen, instead of a pencil, to mark the chromatography paper?

2. Why do you have to cover the beaker while the solvent is moving up the paper?

3. What problem would be caused by moving the beaker during the development of the chromatogram?

4. Make the necessary measurements (in millimeters) and calculate R_F for the following chromatogram:

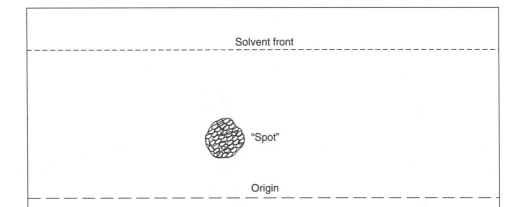

Experiment 3

Work Page

Distance between origin and solvent front: _____ mm

Solutions of Known Ions

Solution	Ion	Color (After spraying)	Distance from Origin (mm)	R_F
A	Fe^{3+}			
B	Cu^{2+}			
C	Co^{2+}			
D	Fe^{3+}			
	Cu^{2+}			
	Co^{2+}			

Solutions of Unknown Ions

Put a check mark in the three right-hand columns to indicate the presence of each respective ion; leave blank if no ion is present. Be sure to enter the identification number of your unknown.

Solution	Unknown No.	Fe^{3+}	Cu^{2+}	Co^{2+}
E				
F				

Experiment 3

Report Sheet

Distance between origin and solvent front: _____ mm

Solutions of Known Ions

Solution	Ion	Color (After spraying)	Distance from Origin (mm)	R_F
A	Fe^{3+}			
B	Cu^{2+}			
C	Co^{2+}			
D	Fe^{3+}			
	Cu^{2+}			
	Co^{2+}			

Solutions of Unknown Ions

Put a check mark in the three right-hand columns to indicate the presence of each respective ion; leave blank if no ion is present. Be sure to enter the identification number of your unknown.

Solution	Unknown No.	Fe^{3+}	Cu^{2+}	Co^{2+}
E				
F				

Densities of Liquids and Solids

Performance Goal

4–1 Calculate the density of a liquid or a solid from experimental data.

CHEMICAL OVERVIEW

One of the physical properties that characterize a substance is its **density,** which is defined as its *mass per unit volume.* Mathematically,

$$\text{Density} = \frac{\text{mass}}{\text{volume}} \tag{4.1}$$

According to this equation, density is equal to the ratio of the mass of a sample of a substance to the volume it occupies. The density of a solid is normally expressed in grams per cubic centimeter (g/cm^3), the density of a liquid in grams per cubic centimeter or grams per milliliter (g/mL), and the density of a gas in grams per liter (g/L).

To determine the density of a substance, you must measure both the mass and volume of the same sample of the substance. Density is then calculated by dividing the mass by the volume, as indicated in Equation 4.1. Mass is measured by the usual weighing techniques. The volume of a liquid may be measured in a graduated cylinder. The dimensions of a solid with a regular geometric shape (rectangular block, cylinder, sphere) may be measured with a ruler, and these measurements can then be used to calculate the volume. The volume of a solid with an irregular shape may be determined by measuring the volume of a liquid displaced when the solid is immersed in the liquid.

In Part 1 of this experiment you will be asked to determine experimentally the density of a known substance and then to calculate the **percent error** in your determination. Percent error is defined by the following equation:

$$\text{Percent error} = \frac{\text{error}}{\text{accepted value}} \times 100 \tag{4.2}$$

The "error" is the difference between the experimental value and the accepted value. Error is expressed as an **absolute value,** i.e., a numerical value without regard for algebraic sign. Absolute value is indicated by enclosing the quantity between vertical lines. Thus Equation 4.2 becomes

$$\text{Percent error} = \frac{|\text{experimental value} - \text{accepted value}|}{\text{accepted value}} \times 100 \qquad (4.3)$$

SAFETY PRECAUTIONS AND DISPOSAL METHODS

Safety hazards in this experiment cannot be identified precisely because of the wide variety of chemicals that might be used as liquid unknowns. This uncertainty dictates that all liquids be regarded as potentially dangerous and treated accordingly. This includes the known liquid, trichloroethane. Liquid samples should be obtained from a dispensing station in the hood. If taken from the hood, liquids should be in containers that are stoppered or covered with a plastic sheet or metal foil. Some unknowns may be flammable; they should therefore be kept away from open flames. When you are finished with them, discard them as directed by your instructor. Avoid contact between all liquids and your skin; if it occurs, wash the exposed area thoroughly with soap and water. *Safety glasses must be worn at all times.*

Depending on the nature of your liquid unknown, disposal directions will be given by your instructor. Trichloroethane should be collected in a stoppered bottle.

PROCEDURE

NOTE: All mass measurements are to be recorded in grams to the nearest 0.01 g or 0.001 g if so instructed. Length measurements are to be recorded in centimeters to the nearest 0.1 cm. Record liquid volume measurements in milliliters to the nearest 0.1 mL.

1. Density of a Liquid

A. Your 50-mL graduated cylinder and a piece of plastic wrap (e.g., Saran wrap) to cover the opening constitute your "container" (see page 10) for Part 1 of this experiment. Being sure the cylinder is clean and dry, weigh it and the Saran wrap—the container—to the nearest 0.01 g on a centigram balance. Record the mass in the proper number of significant figures on your work page.

B. Take the cylinder and plastic wrap to the hood. Pour 12 to 15 mL of 1,1,1-trichloroethane into the cylinder; do *not* attempt to make the amount *exactly* 12, 13, 14, or 15 mL. Cover the cylinder with the plastic wrap. Estimate the volume to the nearest 0.1 mL (see page 9 on reading volume), and record that value to the proper number of significant figures.

C. After making sure the outside of the cylinder is dry, measure and record the mass of the container plus liquid on the centigram balance.

D. Dispose of your liquid as directed by your instructor.

E. In the same manner, collect data for finding the densities of one or more unknown liquids, as required by your instructor. *Be sure to record the identification number of each unknown.*

2. Density of a Regular Solid

Select one or more of the solid unknowns provided for this experiment and record its identification number. Determine and record its mass to the nearest centigram. Make whatever measurements may be necessary to calculate the volume of the object, listing these measurements to the closest 0.1 cm. Because these objects are of various shapes, the data table contains blank spaces in which to describe the shapes and identify the measurements (length, diameter, etc.) that are made.

3. Density of an Irregular Solid

A. Place 20 to 25 mL of water into the cylinder from Part 1. Record the volume to the nearest 0.1 mL. Determine the mass of the cylinder plus water to the nearest centigram. This is the mass of the container for Part 3.

B. Select and record the identification number of one of the unknown irregular solids provided for this experiment. Place enough of the solid into the graduated cylinder to cause the liquid level to rise by more than 10 milliliter markings. Be sure all of the solid is below the surface of the liquid. Record the volume to the nearest 0.1 mL. Also measure the mass of the container and its contents to the nearest centigram.

C. Dispose of your solid material into the recovery facility that has been set up in your laboratory. Be careful not to mix unknown solids.

D. Repeat Steps 3A and 3B for as many unknowns as are required by your instructor, or for a second run with the same unknown.

CALCULATIONS

Be sure to include units in the results of all calculations. Also be sure to express those results in the correct number of significant figures.

1. Density of a Liquid

Find the mass of the liquid by difference—by subtracting the mass of the container from the mass of the container plus liquid. The density of the liquid is found by dividing the mass of the liquid sample by its volume, as indicated in Equation 4.1. Percent error may be calculated by substituting into Equation 4.3; be careful of significant figures in the result. The accepted value for the density of 1,1,1-trichloroethane is 1.34 g/mL.

2. Density of a Regular Solid

The volume of a rectangular solid is calculated by multiplying the length by the width by the height: $V = l \times w \times h$.

The volume of a cylinder is the area of the base times the height. The area of a circle is $\pi d^2/4$, where d is the diameter. Thus

$$V_{\text{cylinder}} = \frac{\pi d^2 h}{4}$$

The value of π to eight decimal places is 3.14159265; the number of places to which you should round it off is left to you.

The volume of a sphere is found from the following equation:

$$V_{\text{sphere}} = \frac{\pi d^3}{6}$$

Once you have calculated the volume of the unknown solid, its density may be found by substituting into Equation 4.1, as before.

3. Density of an Irregular Solid

Both the mass and the volume of the sample are found by difference. Density is again calculated by substitution into Equation 4.1.

Experiment 4

Advance Study Assignment

1. The volume of an unknown liquid is 28.6 mL and its mass is 32.2 grams. Calculate the density of the liquid.

2. When 95.0 g of an unknown metal are submerged in water in a graduated cylinder, the water level rises from 38.2 to 49.5 mL. Calculate the density of the metal.

3. The accepted value for the density of a certain metal is 5.48 g/cm^3. Calculate the percent error in a laboratory experiment that yields a value of 5.2 g/cm^3. Express this result in the proper number of significant figures.

Experiment 4

Work Page

Part 1—Density of a Liquid

Liquid (List identification number of unknowns)	Trichloroethane	Unknown _____	Unknown _____	Unknown _____
Mass of container + liquid (g)				
Mass of container (g)				
Mass of liquid (g)				
Volume of liquid (mL)				
Density (g/mL)				
Percent error (trichloroethane only)				

"Accepted" value for density of 1,1,1-trichloroethane: 1.34 g/mL.

Calculation Setups for Density Determinations:

Calculation Setups for Percent Error for Trichloroethane:

Part 2—Density of a Regular Solid

Unknown Number				
Shape of unknown				
Volume of unknown (cm^3)				
Mass of unknown (g)				
Density (g/cm^3)				

Calculation Setups for Determination of Volumes of Unknowns:

For each unknown, list the measurements taken and show calculation setup.

Calculation Setups for Density Determinations:

Experiment 4

Work Page

Part 3—Density of an Irregular Solid

Unknown Number				
Mass of container + liquid + solid (g)				
Mass of container + liquid (g)				
Mass of solid (g)				
Volume of liquid + solid (mL)				
Volume of liquid (mL)				
Volume of solid (cm^3)				
Density (g/cm^3)				

Calculation Setups for Density Determinations:

Experiment 4

Report Sheet

Part 1—Density of a Liquid

Liquid (List identification number of unknowns)	Trichloroethane	Unknown _____	Unknown _____	Unknown _____
Mass of container + liquid (g)				
Mass of container (g)				
Mass of liquid (g)				
Volume of liquid (mL)				
Density (g/mL)				
Percent error (trichloroethane only)				

"Accepted" value for density of 1,1,1-trichloroethane: 1.34 g/mL.

Calculation Setups for Density Determinations:

Calculation Setups for Percent Error for Trichloroethane:

Part 2—Density of a Regular Solid

Unknown Number				
Shape of unknown				
Volume of unknown (cm^3)				
Mass of unknown (g)				
Density (g/cm^3)				

Calculation Setups for Determination of Volumes of Unknowns:

For each unknown, list the measurements taken and show calculation setup.

Calculation Setups for Density Determinations:

Experiment 4

Report Sheet

Part 3—Density of an Irregular Solid

Unknown Number				
Mass of container + liquid + solid (g)				
Mass of container + liquid (g)				
Mass of solid (g)				
Volume of liquid + solid (mL)				
Volume of liquid (mL)				
Volume of solid (cm^3)				
Density (g/cm^3)				

Calculation Setups for Density Determinations:

Simplest Formula of a Compound

Performance Goals

5–1 Prepare a compound and collect data from which you can determine the mass of each element in the compound.

5–2 From the mass of each element in a compound, determine its simplest formula.

CHEMICAL OVERVIEW

Chemical compounds are composed of atoms of different elements. The atoms are held together by chemical bonds. It has been shown experimentally that the ratio of moles of the elements in a compound is nearly always a ratio of small, whole numbers. The few exceptions are known as nonstoichiometric compounds. The formula containing the lowest possible ratio is known as its **simplest formula.** It is also called the **empirical formula.** At times it may be the same as the molecular formula; often, however, the molecular formula is an integral multiple of the simplest, empirical formula. For example, the simplest formula of the compound benzene (C_6H_6) is simply CH, indicating that the ratio of carbon atoms to hydrogen atoms is one to one.

To find the simplest formula of a compound, you will combine the elements in the compound under conditions that will allow you to determine the mass of each element. From these data the moles of atoms of each element may be calculated. By dividing these numbers by the smallest number of moles, you obtain quotients that are in a simple ratio of integers, or are readily converted to such a ratio. The ratio of moles of atoms of the elements in a compound is the same as the ratio of individual atoms that is expressed in the empirical formula.

Remember: The essential information required to find the simplest formula of a compound is the number of grams of each element in a sample of the compound.

In Option 1 you will react a measured mass of copper with excess sulfur. The excess sulfur is burned away as sulfur dioxide. In Option 2 the reaction is between a measured quantity of tin and excess nitric acid. The excess acid is boiled off. Option 3 involves the reaction of a measured mass of magnesium with excess oxygen from the air.

Your instructor may require you to perform the experiment twice to obtain duplicate results, or to complete more than one option. If so, plan your use of time. The procedure includes some periods in which you wait for a crucible to cool. The cooling periods in the first run of the experiment can be used for heating periods in the second run, and vice versa.

SAMPLE CALCULATIONS

A piece of aluminum is ignited in a suitable container, yielding an oxide. Calculate the simplest formula of the oxide from the following data:

Mass of container	17.84 g
Mass of container + aluminum	18.38 g
Mass of container + compound	18.86 g

1. Mass of each element from data:

Mass of container + aluminum		mass of container		mass of aluminum
18.38 g	−	17.84 g	=	0.54 g aluminum

Mass of container + compound		mass of container + aluminum		mass of oxygen
18.86 g	−	18.38 g	=	0.48 g oxygen

2. Moles of each element:

$$0.54 \text{ g Al} \times \frac{1 \text{ mole Al atoms}}{27.0 \text{ g Al}} = 0.020 \text{ mole Al atoms}$$

$$0.48 \text{ g O} \times \frac{1 \text{ mole O atoms}}{16.0 \text{ g O}} = 0.030 \text{ mole O atoms}$$

3. Simplest formula ratio:

Obtain the ratio of atoms by dividing the number of moles of each atom by the smallest number of moles:

$$\text{Al:} \quad \frac{0.020}{0.020} = 1.0; \qquad \text{O:} \quad \frac{0.030}{0.020} = 1.5$$

The ratio is 1.0 mole aluminum atoms to 1.5 moles oxygen atoms. Change this ratio to a whole number ratio by multiplying each value by 2:

$$\text{Moles of Al atoms} = 1.0 \times 2 = 2.0$$

$$\text{Moles of O atoms} = 1.5 \times 2 = 3.0$$

The simplest formula is therefore Al_2O_3.

Many students find it convenient to organize their calculations by arranging both data and results in a table as follows:

Element	Grams	Moles	Mole Ratio	Formula Ratio
Al	0.54	0.020	1.0	2
O	0.48	0.030	1.5	3

SAFETY PRECAUTIONS AND DISPOSAL METHODS

The safety considerations in this experiment relate to the operation of a Bunsen burner and the handling of hot items. Blue burner flames are visible, but easily lost against a laboratory background. Be careful not to reach through one in reaching for some object behind it. If you have long hair, tie it back so it does not get into the flame. Be sure to use crucible tongs when handling hot crucibles, including the lid. Laboratory hardware gets hot, too. Harmful gases are released in Options 1 and 2. These reactions must be performed in a fume hood, as stated in their procedures. Be careful of hot chemical spattering from crucibles when they are heated. Be sure to wear goggles throughout this experiment.

Dispose of any solid residue as directed by your instructor.

PROCEDURE

Option 1: A Sulfide of Copper

NOTE: All mass measurements in Option 1 are to be recorded in grams to the nearest 0.01 g or 0.001 g if so instructed.

A. The purpose of this step is to remove moisture from the crucible. Support a clean, dry porcelain crucible and its lid on a clay triangle, as shown in Figure 5.1. Heat it slowly at first, and then fairly strongly in the direct flame of a burner for about 4 to 5 minutes. Set the crucible and lid aside on a wire gauze to cool.

B. When the crucible and lid are cool to the touch, weigh them on a centigram balance. Record this value as the mass of the container.

C. Place a loosely rolled ball of copper wire or medium shavings, about 1.5 to 2 g, into the crucible. Weigh them, with the lid, on a centigram or milligram balance, and record the weight as the mass of the container plus metal.

D. Sprinkle about 1 to 1.5 g of powdered sulfur over the copper. Place the lid on the crucible and begin heating it in a fume hood. Heat slowly at first, and then with a moderate flame until the sulfur no longer burns around the lid. Finally, heat the crucible strongly for about 5 minutes, making sure that no excess sulfur is present on the lid or on the sides of the crucible. It sometimes helps to hold the burner at its base and direct the flame under the lip of the lid all the way around.

E. Set the container and its contents aside to cool. Do not open the lid until the crucible is cool, because air oxidation is apt to occur.

F. When the crucible is cool, lift the lid and examine the contents. There should be no evidence of sulfur in the crucible or on the lid. If sulfur is present, heat the crucible again until the sulfur is completely burned away. Allow the crucible to cool.

G. Weigh the container and its contents again. Record your measurement as the mass of the container plus compound.

H. Set the container aside while you complete your calculations. Do not discard your compound until your calculations are finished and satisfactory; if they are not satisfactory, it is possible you may be able to salvage your work if the material is still on hand.

I. Just before discarding the compound, press it to the bottom of the crucible. Notice the difference between the physical properties of the compound and those of the elements from which it was formed. The compound should be discarded as directed by your instructor.

Figure 5.1
Heating of a porcelain crucible

Option 2: An Oxide of Tin

NOTE: All mass measurements in Option 2 are to be recorded in grams to the nearest 0.01 g. or 0.001 g if so instructed.

A. Heat a porcelain crucible and its lid as described in Option 1, Step A. Allow it to cool.

B. Weigh as described in Option 1, Step B.

C. Place a loosely rolled ball of tin foil, weighing 1 to 1.5 g, into the crucible. Weigh the crucible, the lid, and the metal on a centigram balance. Record the mass obtained.

D. Under the fume hood, add concentrated nitric acid, HNO_3, drop by drop, to the crucible until all the tin has reacted and a damp white paste remains.

E. Heat the paste cautiously with a mild flame, taking care not to cause spattering. After all of the liquid has evaporated, heat the crucible with a hot flame for 5 minutes.

F. Cool the crucible and compound to room temperature and weigh it. Record the mass of the container and compound.

G. Keep your compound in the crucible until all calculations are completed. This may save you time if it becomes necessary to add more nitric acid.

Option 3: Magnesium Oxide

NOTE: All mass measurements in Option 3 are to be recorded in grams to the nearest 0.001 g.

A. Heat a porcelain crucible and its lid as described in Option 1, Step A. Allow it to cool.

B. Weigh the crucible and lid on a milligram balance. Record this value on the work page as mass of the container.

C. Place a loosely folded magnesium ribbon, weighing 0.5 to 0.7 g, into the crucible. Weigh the crucible, lid, and metal on a milligram balance and record the mass.

D. Remove the lid and hold it near the crucible with a pair of tongs. Start heating the crucible, and as soon as the magnesium begins to burn, replace the lid. Continue the process, holding the escape of white smoke to a minimum (very finely divided magnesium oxide looks like smoke). When the contents of the crucible no longer burn, cock the lid wide enough to allow a sufficient amount of air to enter to complete the reaction, as shown in Figure 5.1, and heat it strongly for 5 minutes.

E. To convert the possible side product, magnesium nitride, to the oxide, let the crucible cool, add 10 drops of deionized water to it, and then gently heat to vaporize excess water. CAUTION: SPATTERING MAY OCCUR.

F. Finish heating the crucible with a strong flame for 5 to 8 minutes.

G. Allow the crucible to cool, then weigh the cool crucible, lid, and product on a milligram balance.

Experiment 5

Advance Study Assignment

1. Circle one of the following formulas that is correctly written as an empirical formula:

$$NaSO_{1.5} \qquad (H_2NO)_2 \qquad Fe_3O_4 \qquad N_2O_4$$

2. 6.25 grams of pure iron are allowed to react with oxygen to form an oxide. If the product weighs 14.31 grams, find the simplest formula of the compound.

3. In determining the simplest formula of lead sulfide, 2.46 grams of lead are placed in a crucible with 2.00 grams of sulfur. When the reaction is complete, the product has a mass of 3.22 grams. What mass of sulfur should be used in the simplest formula calculation? Find the simplest formula of lead sulfide.

Experiment 5

Work Page

Data

Option or Trial Number				
Mass of container (g)				
Mass of container + metal (g)				
Mass of container + compound (g)				

Results for Sample of Compound Prepared

Mass of metal (g)				
Mass of nonmetal element (g)				
Moles of metal				
Moles of nonmetal element				
Ratio: $\dfrac{\text{moles metal}}{\text{moles nonmetal}}$ *	$\overline{1}$	$\overline{1}$	$\overline{1}$	$\overline{1}$
Simplest formula				

*Express this ratio as a decimal number over $1 \left(\text{e.g.,} \frac{3.044}{1}\right)$.

Show calculations on the reverse side of this page.

Calculations

Experiment 5

Report Sheet

Data

Option or Trial Number				
Mass of container (g)				
Mass of container + metal (g)				
Mass of container + compound (g)				

Results for Sample of Compound Prepared

Mass of metal (g)				
Mass of nonmetal element (g)				
Moles of metal				
Moles of nonmetal element				
ratio: $\dfrac{\text{moles metal}}{\text{moles nonmetal}}$ *	$\overline{}$ 1	$\overline{}$ 1	$\overline{}$ 1	$\overline{}$ 1
Simplest formula				

*Express this ratio as a decimal number over 1 (e.g., $\frac{3.044}{1}$).

Show calculations on the reverse side of this page.

Calculations

Hydrates

Performance Goals

6–1 Calculate the percentage of water of hydration in a compound from experimental data.

6–2 Calculate the formula of a hydrate of a known anhydrous salt from experimental data.

CHEMICAL OVERVIEW

Hydrates are chemical compounds that contain water as part of their crystal structure. This water is quite strongly bound and is present in a definite proportion relative to other constituents. It is referred to as **water of hydration.**

The formula of a hydrate consists of the formula of the **anhydrous** (without water) **compound** followed by a dot, the number of water molecules that crystallize with one formula unit of the compound, and the formula of water. For example, $CuSO_4 \cdot 5H_2O$ indicates that 5 molecules of water—H_2O—crystallize with 1 formula unit of anhydrous copper(II) sulfate, $CuSO_4$, to form copper sulfate pentahydrate.

Generally, water of hydration can be driven from hydrates by heating, leaving behind the anhydrous salt. The process may be accompanied by physical changes, such as a change in color or physical appearance. For example, $CuSO_4 \cdot 5H_2O$ is an intense blue shiny crystal which, upon heating, turns into pale green-blue powdery anhydrous $CuSO_4$.

In this experiment you will be instructed to determine the mass of a sample of an unknown hydrate by difference, using a preweighed crucible as the container. The substance will be "dehydrated" by heat and weighed again. The loss of mass represents the mass of water in the original sample, which may be expressed as percentage of water of hydration, using the usual (part quantity/total quantity) × 100 relationship:

$$\text{Percentage water of hydration} = \frac{\text{grams of water}}{\text{grams of hydrate}} \times 100 \qquad (6.1)$$

To find the formula of the original hydrate, you will determine from the preceding data the mass of the anhydrous compound, the formula of which will be given to you. From this you can calculate the moles of anhydrous compound in the original sample. From the mass of water in the original sample you can calculate the moles of water. By dividing the moles of water by the moles of anhydrous salt, you obtain the ratio of moles of water to 1 mole of anhydrous salt, as in the 5-to-1 ratio for $CuSO_4 \cdot 5H_2O$ above.

In this experiment you will be directed to "heat to constant mass." Your purpose is to heat the substance until *all* of the water is driven off. After a first heating, cooling, and weighing, you cannot tell if all water has been removed or if some still remains. You therefore repeat the heating, cooling, and weighing procedure. If the same mass is reached after the second heating, you may assume that all water was removed the first time. If mass was lost in the second heating, you may be sure that all water was *not* removed in the first heating, and you are still unsure that it was all driven off in the second heating. Another heating is therefore required. The heating, cooling, and weighing sequence is repeated until two successive duplicate weighings are recorded. Two weighings within the ± uncertainty range of the balance are generally considered to be duplicate weighings; duplication within 0.005 g is satisfactory for this experiment.

Your instructor may require you to perform the experiment twice to obtain duplicate results or to run more than one unknown. If so, plan your use of time. The procedure includes some periods in which you must wait for a crucible to cool. The cooling periods in the first run of the experiment can be used for heating periods in the second run and vice versa. In this way you perform both runs simultaneously.

SAFETY PRECAUTIONS AND DISPOSAL METHODS

The safety considerations in this experiment relate to the operation of a Bunsen burner and the handling of hot items. Blue burner flames are visible, but easily lost against a laboratory background. Be careful not to reach through a flame in picking up some object behind it. If you have long hair, tie it back so it does not get into the flame. Be sure to use crucible tongs in handling hot crucibles, including the lids. Laboratory hardware gets hot, too. Goggles should always be worn when working with chemicals, and particularly while heating them, as in this experiment. Be alert also to the possibility of hot chemicals "shooting" out of the test tube in Part 2.

The anhydrous solids should be discarded in a container or you should follow directions given by your instructor.

PROCEDURE

NOTE: Record all mass measurements in grams to the nearest 0.001 g.

1. Percentage Water in a Hydrate: Formula of a Hydrate

A. Heat a clean porcelain crucible and its lid—the "container" for this experiment (see pages 9 and 10)—on a clay triangle over a direct flame for 5 minutes to drive off any surface moisture. (See Figure 5.1, page 62.) When they are cool to the touch, weigh them on a milligram balance, recording this and subsequent masses in the "constant mass" portion of your data sheet. Heat the container to constant mass in 5-minute heating cycles until duplicate masses (within 0.005 g) are reached. Record the final (constant) mass as the mass of the container.

NOTE: Part 2 of the experiment may be performed during the cooling cycles of Part **1**.

B. Place 1 to 1.5 g of the unknown solid hydrate into the container, and weigh again on a milligram balance. Record as the mass of the container plus hydrate.

C. With the lid almost covering the crucible, heat the container and its contents, gently at first, and then with a hot flame for 10 minutes. Cool the crucible and lid, and when they feel cool to the touch, weigh them again. Record this and subsequent weighings in the "Constant Mass Data" portion of the work page. Heat to constant mass in 4- to 6-minute heating cycles until duplicate masses are reached. The final (constant) mass should be recorded as the mass of the container plus anhydrous salt.

D. Set the container and its contents aside while you complete your calculations. Do not discard the residue until your calculations are finished and satisfactory; if they are not satisfactory, it is possible that you may be able to salvage your work if the material is still on hand.

2. Behavior of a Hydrate

A. Place a few small crystals of $CoCl_2 \cdot 6H_2O$ into a test tube. Holding the test tube tilted at an angle, and with its mouth pointing away from you and all others, heat the test tube gently. Record your observations.

B. After the test tube has cooled to room temperature, add a few drops of water. Hold the test tube against the back of your hand. Record your observations.

CALCULATIONS

From the masses of the container, the container plus hydrate, and the container plus anhydrous salt, identify and perform the subtractions that will yield the mass of the hydrate and the mass of the water of hydration. From these, calculate the percentage water of hydration, using Equation 6.1.

Return to the original data (masses of container, container plus hydrate, and container plus anhydrous salt), and identify and perform the subtraction that will yield the mass of the anhydrous salt. The anhydrous salts resulting from dehydration of the different unknowns in this experiment are:

Unknown	*yields*	*Anhydrous Salt*
A		$BaCl_2$
B		$MgSO_4$
C		$CuSO_4$

Convert the grams of anhydrous salt to moles. Also convert grams of water of hydration to moles. Calculate the number of moles of water per mole of anhydrous compound; enter this result in the "Ratio" line in your table of results. From the ratio, determine the formula of the hydrate.

Experiment 6

Advance Study Assignment

1. How can you make sure that all of the water of hydration has been removed? Explain.

2. Why do you have to cool the crucible before weighing it?

3. A 2.815-g sample of $CaSO_4 \cdot XH_2O$ was heated until all of the water was removed. Calculate the percentage water of hydration and the formula of the hydrate if the residue after heating weighed 2.485 g.

Experiment 6

Work Page

Data and result tables for Experiment 6 appear on pages 78 and 79.

Part 1—Percentage Water in a Hydrate; Formula of a Hydrate

Constant Mass Data (Supply identification letters or numbers for unknowns)

Item	Container	Unknown _____	Unknown _____	Unknown _____	Unknown _____
1st heating (g)					
2nd heating (g)					
3rd heating (g)					
4th heating (g)					
Final heating (g)					

Mass Data

Container (g)				
Container + hydrate (g)				
Container + anhydrous salt (g)				

Results

(Show full calculations for one column at the top of the next page)

Mass of hydrate (g)				
Mass of water of hydration (g)				
Percentage water of hydration				
Mass of anhydrous salt (g)				
Formula of anhydrous salt				
Moles of anhydrous salt				
Moles of water of hydration				
Ratio: $\dfrac{\text{moles of water}}{\text{1 mole of anhydrous salt}}$ *	$\dfrac{\quad}{1}$	$\dfrac{\quad}{1}$	$\dfrac{\quad}{1}$	$\dfrac{\quad}{1}$
Formula of hydrate				

*Express this ratio as a decimal number over 1 (e.g., $\frac{2.875}{1}$).

Experiment 6

Work Page

Calculations for One Full Column of Results from Previous Page

Part 2—Behavior of a Hydrate

2A. Observations when a hydrate is heated in a test tube.

2B. Observations when water is added to an anhydrous salt.

Experiment 6

Report Sheet

Data and result tables for Experiment 6 appear on pages 82 and 83

Part 1—Percentage Water in a Hydrate; Formula of a Hydrate

Constant Mass Data (Supply identification letters or numbers for unknowns)

Item	Container	Unknown _____	Unknown _____	Unknown _____	Unknown _____
1st heating (g)					
2nd heating (g)					
3rd heating (g)					
4th heating (g)					
Final heating (g)					

Mass Data

Container (g)				
Container + hydrate (g)				
Container + anhydrous salt (g)				

Results

(Show full calculations for one column at the top of the next page)

Mass of hydrate (g)				
Mass of water of hydration (g)				
Percentage water of hydration				
Mass of anhydrous salt (g)				
Formula of anhydrous salt				
Moles of anhydrous salt				
Moles of water of hydration				
Ratio: $\dfrac{\text{moles of water}}{\text{1 mole of anhydrous salt}}$ *	$\dfrac{\quad}{1}$	$\dfrac{\quad}{1}$	$\dfrac{\quad}{1}$	$\dfrac{\quad}{1}$
Formula of hydrate				

*Express this ratio as a decimal number over 1 (e.g., $\frac{2.875}{1}$).

Experiment 6

Report Sheet

Calculations for One Full Column of Results from Previous Page

Part 2—Behavior of a Hydrate

2A. Observations when a hydrate is heated in a test tube.

2B. Observations when water is added to an anhydrous salt.

Percentage of Oxygen in Potassium Chlorate

Performance Goal

7–1 Determine the percentage of one part of a compound from experimental data.

CHEMICAL OVERVIEW

The thermal decomposition of potassium chlorate is described by the equation

$$2\ KClO_3(s) \xrightarrow{\ \Delta\ } 2\ KCl(s) + 3\ O_2(g) \tag{7.1}$$

In this experiment you will determine the percentage of oxygen in potassium chlorate. You will compare your experimental result with the theoretical percentage calculated from the formula $KClO_3$.

While potassium chlorate decomposes simply by heating, the reaction is intolerably slow. A catalyst, manganese dioxide, MnO_2, is therefore added to speed the reaction. Although it contains oxygen, the catalyst experiences no permanent change during the reaction and does not contribute measurably to the amount of oxygen generated. As with all catalysts, the quantity present at the end of the reaction is the same as the quantity at the beginning.

The experimental procedure is to weigh a quantity of potassium chlorate, heat it to drive off the oxygen, and then weigh the residue, which is assumed to be potassium chloride. The loss in mass represents the oxygen content of the original potassium chlorate.

The above procedure will be carried out in a test tube. The "container" (see page 10) for this experiment will be more than just the test tube, however. It will include the constant mass of catalyst that remains in the test tube throughout the experiment, plus whatever device is used to hold the test tube and its contents while they are being weighed. If your milligram balance has a suspended pan and there is provision for hanging an object to be weighed, you can clamp the test tube in a test-tube holder and hang the entire assembly from the hook provided. In this case the test-tube holder is a part of the container. If the pan on your milligram balance is supported only from beneath, you can stand your test tube in a small beaker each time it is weighed, and include the beaker in the mass of the container. You must be sure, of course, to use the same beaker for each weighing.

In a thermal decomposition such as this, the product must be "heated to constant mass" before you can be sure the decomposition is complete. After the first heating, cooling, and weighing of the decomposed product, the test tube is heated, cooled, and weighed again. If the two weighings are the same, within the plus-and-minus uncertainty of the equipment used, it may be assumed that all of the oxygen was removed in the first heating. If mass is lost in the second heating, it means that some oxygen remained after the first heating and was driven off in the second. It is possible that some oxygen may still be present after the second heating, too. The procedure is therefore repeated again, as many times as necessary, until there is negligible change in mass (no more than 0.005 g for this experiment) between two consecutive weighings.

SAFETY PRECAUTIONS AND DISPOSAL METHODS

CAUTION

This experiment is potentially hazardous, and if performed carelessly could lead to a serious accident!

The formation of a gas at the bottom of a test tube may result in a sudden expansion, blowing hot chemicals out of the test tube. This will not occur if the test tube is handled properly during heating. When heating a solid in a test tube, tip the tube until it is almost horizontal and tap it carefully until the contents are distributed over the lower half of the length of the tube, as shown in Figure 7.1. Holding it at about this angle, move the test tube back and forth in the flame of the burner, distributing the heat over the entire length of the mixture. *Do not concentrate the heat in any one area,* **particularly near the bottom of the test tube.**

Be very sure your test tube is not pointing toward anybody, including yourself, while it is being heated. Be aware of what those around you are doing while this experiment is being performed in the laboratory. Do not place yourself in the firing line of somebody else's test tube, and if another student points a test tube toward your work station, ask him or her to point it elsewhere. Obviously, *wearing goggles is absolutely mandatory while you or anyone near you is performing this experiment.*

Dispose of the residue in the designated container.

Figure 7.1

$KClO_3 + MnO_2$ mixture

Heating range of test tube

Test-tube clamp here

PROCEDURE

NOTE: Record all mass measurements in grams to the nearest 0.001 g.

A. Place 0.5 to 0.8 g of manganese dioxide, MnO_2, into a large-size test tube. (Do not use a *small* test tube; they are more inclined to "shoot" their contents.) Heat the test tube over a Bunsen burner for about 3 to 4 minutes to drive off any moisture that may be present in the catalyst and test tube. When the tube is cool to the touch, measure the mass of the entire container on a milligram balance. Record the mass in the space provided.

B. Add about 1.0 to 1.5 g of potassium chlorate, $KClO_3$, to the test tube. Find and record the mass of the container and its contents.

C. Mix the contents of the test tube until they have a somewhat uniform gray appearance. (Be careful not to lose any of the contents.) Carefully observing the procedure and safety precaution above, heat the test tube and its contents. Heat gently at first, increasing the intensity after the mixture seems to "boil," as it sometimes appears to do when bubbles of oxygen are being released. Continue heating for about 5 minutes, and then cool and weigh. Repeat the process in 3- to 5-minute heating cycles until constant mass is reached.

D. Set the container and its contents aside while you complete your calculations. Do not discard the residue until your calculations are finished and satisfactory; if they are not satisfactory, it is possible that you may be able to salvage your work if the material is still on hand.

CALCULATIONS

Using only numbers in the data portions of your work page, calculate by difference the initial mass of potassium chlorate and the mass of oxygen released in heating. From these quantities, find the experimental percentage of oxygen in potassium chlorate. Calculate the theoretical oxygen percentage *from the formula of the compound.* Using the theoretical percentage as the accepted value, calculate your percentage error by the equation

$$\text{Percentage error} = \frac{\text{error}}{\text{accepted value}} \times 100$$

$$= \frac{|\text{experimental value} - \text{accepted value}|}{\text{accepted value}} \times 100$$

Note that the numerator of this equation is an absolute value, simply the difference between the observed and theoretical values expressed as a positive quantity.

Experiment 7

Advance Study Assignment

1. What potential source of laboratory accident is present in this experiment? Explain the procedure you will follow to minimize this hazard. What two precautions will you follow to prevent injury to yourself, and what will you do to avoid injuring another in the event the accident does occur?

2. Another thermal decomposition that produces oxygen begins with silver oxide: $2\,Ag_2O \longrightarrow 4\,Ag + O_2$. In a hypothetical experiment, a student collects the following data:

Mass of crucible	30.296 g
Mass of crucible + Ag_2O	38.623 g
Mass of crucible + contents after	
"complete" decomposition	38.061 g

Calculate the following:

a. The starting mass of Ag_2O:

b. The mass of oxygen released:

c. Experimental percentage oxygen in Ag_2O *from data:*

d. Theoretical percentage oxygen in Ag_2O *from the formula:*

e. Percentage error:

Experiment 7

Work Page

Constant Mass Data

Trial	1	2	3	4
1st heating (g)				
2nd heating (g)				
3rd heating (g)				
4th heating (g)				
Final heating (g)				

Mass Data

Mass of container (g)				
Mass of container + $KClO_3$ (g)				
Mass of container + KCl (g)				

Results

(Show all calculations for one column on the next page)

Mass of $KClO_3$ used (g)				
Mass of oxygen released (g)				
Percentage oxygen				
Percentage error				

Complete Calculations from One Column of Table:

Calculation of Percent Oxygen in KClO$_3$ from Formula:

Experiment 7

Report Sheet

Constant Mass Data

Trial	1	2	3	4
1st heating (g)				
2nd heating (g)				
3rd heating (g)				
4th heating (g)				
Final heating (g)				

Mass Data

Mass of container (g)				
Mass of container + $KClO_3$ (g)				
Mass of container + KCl (g)				

Results

(Show all calculations for one column on the next page)

Mass of $KClO_3$ used (g)				
Mass of oxygen released (g)				
Percentage of oxygen				
Percentage error				

Complete Calculations from One Column of Table:

Calculation of Percent Oxygen in KClO₃ from Formula:

Calorimetry

Performance Goals

8–1 Calculate the specific heat of an unknown solid element by measuring the heat exchanged in a calorimeter.

CHEMICAL OVERVIEW

Heat, a form of energy, can be gained or lost by an object. When the object cools, it loses heat energy; when it is heated, it gains energy. The unit in which heat is measured is the **joule, J.** A joule is a derived unit, having base units of $kg \cdot m^2/sec^2$. The joule is a very small amount of heat, so the **kilojoule, kJ,** is commonly used.[*]

Heat flow is the change in "heat content" of an object as heat energy passes between the object and its surroundings. It is proportional to the mass of the object and its change in temperature. The proportional relationship becomes an equation if the proportionality constant called **specific heat** is introduced:

$$Q = (\text{mass})(\text{specific heat})(\Delta T)$$
$$= m \times c \times \Delta T \qquad (8.1)$$

where Q is the heat flow in joules, m is mass in grams, c is specific heat in joules per gram degree, or $J/g \cdot {}^\circ C$, and ΔT is the temperature change, or final temperature minus initial temperature. The Greek letter delta, Δ, indicates a change in a measured value and always means final value minus initial value: $\Delta X = (X_f - X_i)$.

Specific heat is a property of a pure substance. It is the number of joules of energy that are required to raise the temperature of 1 g of the substance by 1°C. The specific heat of water is $4.18 \ J/g \cdot {}^\circ C$. This value is used often in calorimetry experiments; it is one you should remember.

When a "hot" object comes into contact with a "cold" object—that is, when an object at higher temperature comes into contact with an object at lower temperature—heat flows from the hot object to the cold object. The hot object raises the temperature of the cold object, and the cold object cools the hot object. Eventually they reach the same intermediate temperature.

---

[*]The joule is still in the process of replacing the calorie as the standard heat unit in chemistry, and the calorie is still in common use. The calorie is now defined as exactly 4.184 joules.

Heat flow between objects can be measured in a **calorimeter.** A perfect calorimeter is an isolated segment of the universe that allows no heat to flow to or from its contents during an experiment. It follows from the law of conservation of energy that, in a perfect calorimeter,

$$\sum Q = 0 \tag{8.2}$$

where $\sum Q$ is the sum of all individual changes in heat content within the calorimeter. We will assume that the calorimeters used in this experiment are "perfect." This means that any heat transferred between its contents and the surroundings or between its contents and the calorimeter itself are negligible and may be disregarded.

In this experiment, you will measure the specific heat of two metals, one known and the other unknown. The known metal will be copper, which has a specific heat of 0.38 J/g · °C. You will calculate the percent error from this known value. The same experimental method will be used to find the specific heat of the unknown metal.

The laboratory procedure is to heat a weighed metal sample in boiling water until it reaches the temperature of the water. This is the initial temperature of the metal. The mass and initial temperature of water in a "coffee cup" calorimeter are measured. The metal is placed into the calorimeter, and the final temperature reached by both is recorded. According to Equation 8.2, the heat flow of the metal plus the heat flow of the water add up to zero:

$$Q_w + Q_m = 0 \tag{8.3}$$

$$Q_w = -Q_m \tag{8.4}$$

What Equation 8.4 says is that the heat lost by the metal is gained by the water. Substituting the expressions of Equation 8.1 into Equation 8.4 gives

$$m_w \times c_w \times \Delta T_w = -(m_m \times c_m \times \Delta T_m) \tag{8.5}$$

All values are known except the specific heat of the metal. Note the negative sign on the right side of the equation.

SAFETY PRECAUTIONS

No hazardous chemicals are used in this experiment—only hazardous temperatures. You will be working with fairly large quantities of boiling water, pouring from one container to another, and dropping pieces of metal into boiling water. Do it carefully, anticipating and avoiding the conditions that might cause you to lose control of a beaker and burn yourself or others as a result. Also in this experiment you will be working with Bunsen burners. If you have long hair, be sure to tie it back.

PROCEDURE

NOTE: Record all mass measurements in grams to the nearest 0.01 g. Record all temperature measurements in degrees Celsius to the nearest 0.1°C. The smallest graduation on your thermometer must be no larger than 0.2°C for satisfactory results on this experiment.

1. Preparation for Calorimeter Runs

A. Place about 500 mL of tap water into a beaker. Set it aside so it will come to room temperature. *Do not place it near an operating Bunsen burner.* This will be your source of calorimeter water throughout the experiment.

B. Fill a beaker, 600 mL or larger, to about 80% of capacity with deionized water. (The deionized water is not required from a purity standpoint, but its use avoids the buildup of hard-to-remove scale that forms when tap water is boiled in a glass vessel.) Mount the beaker over wire gauze on a ring stand or, if available, use a hot plate and heat the water to boiling. Proceed to Parts lC and 1D while waiting.

  From time to time in the experiment you will have to replace the water that has boiled out of the beaker with other water *already at the boiling temperature.* If some "community" source of boiling water is not available, establish your own by setting up a second beaker in the same manner as the first.

C. Select a piece of copper and an unknown metal sample from among those provided. Record the identification number of the unknown on your data sheet. Measure the mass of each piece and record it in grams to the nearest centigram.

D. Using crucible tongs, place each piece of metal in the water being heated. Position them in such a way that they will be easy to pick up and transfer to the calorimeter later. Be careful not to drop the metal in such a manner that it hits the bottom with sufficient force to break the beaker. The metal piece should remain in the boiling water for at least 30 minutes. As you proceed with the next steps, keep watch over the water level, which must be high enough to cover the metal completely. If too much water boils away, replenish it from the community source of boiling water, or from your second beaker, as explained in Part 1B.

E. When the water in the heating bath is boiling, determine and record its temperature to the nearest 0.1°C. Set the thermometer aside and allow it to cool to room temperature. Two assumptions are made about the temperature of the boiling water: first, that it will remain constant throughout the experiment; and second, that the metal immersed in it will reach the same temperature, which is the initial temperature of the metal for later calculations.

F. Select and prepare a place on the laboratory desk for your calorimeter runs. It should be as far as possible from an operating Bunsen burner from which it might absorb heat, but not so far that the metal loses temperature while being transferred from the boiling water to the calorimeter. It is convenient to place a split rubber stopper on your thermometer (see Figure LP-6 in the Laboratory Procedures section) and clamp it on a ring stand, or to place your thermometer in a buret clamp, as depicted in Figure 8.1.

Figure 8.1

A   B

2. First Calorimeter Run

    A. Measure exactly 100 mL of your calorimeter water in a graduated cylinder and carefully transfer it into your calorimeter. Assuming the density of water to be 1.00 g/mL, the mass of the water is then 100 grams.

    B. Place the thermometer into the calorimeter water, as shown in Figure 8.1. When the temperature has remained constant for about 1 minute, record that temperature to the nearest 0.1°C. In this and all other temperature measurements, be sure the thermometer bulb is totally immersed in the water.

    C. *(The next step is critical. The period in which the metal is out of water must be held to an absolute minimum to keep heat loss to the air as low as possible.)* Using crucible tongs, lift the piece of copper from the boiling water and hold it above the water level but below the top of the beaker for about 2 seconds to allow the boiling water to drain off. Then, *as quickly as possible,* place the metal into the calorimeter water. Be sure it is completely submerged. Be careful not to splash water out of the calorimeter. If this does occur, return the metal to the boiling water and start again at Step 2A.

    D. Continuously and gently stir the water with a glass stirring rod, being careful not to break the bulb of the thermometer. Watch the thermometer. Read and record to the nearest 0.1 °C the highest temperature that is reached before it begins to drop again. This may take several minutes.

    E. Pour the water from the calorimeter into the sink. Place the copper back into the boiling water to be heated for a second run.

3. Subsequent Calorimeter Runs

Again pour exactly 100 mL of water into your calorimeter. Repeat Steps 2B through 2E, using the unknown metal. The metal should be used alternately in subsequent runs, giving each sample enough time in the boiling water to reach its temperature. Make two runs with your copper sample and two with the unknown or as directed by your instructor. At the end of the experiment, dry the metal pieces and return them to the place indicated by your instructor.

CALCULATIONS

Calculate the specific heat of the metal for each run of the experiment, using Equation 8.5, in which the specific heat is the only unknown. Find the percent error for each run with the copper, using the equations

$$\text{Percentage error} = \frac{\text{error}}{\text{accepted value}} \times 100$$

$$\text{Percentage error} = \frac{|\text{experimental value} - \text{accepted value}|}{\text{accepted value}} \times 100$$

Find the average specific heat for the two runs with the unknown metal.

Experiment 8

Advance Study Assignment

1. A. Define specific heat.

   B. Why do you have to transfer the copper rapidly from the boiling water into the calorimeter?

   C. What error would be committed if the boiling water did not cover the metal completely?

2. A student places 1.38 g of unknown metal at 99.6°C into 60.50 g of water at 22.1°C. The entire system reaches a uniform temperature at 31.6°C. Calculate the specific heat of the metal.

3. If the actual specific heat of the metal in Problem 2 is 0.25 J/g·°C, calculate the percentage error.

Experiment 8

Work Page

Data

Run number					
Metal or unknown number					
Mass of metal (g)					
Temperature of boiling water (°C)					
Initial temperature of calorimeter water (°C)					
Final temperature (°C)					

Calculations and Results

Record all results in the table on the following page. Show below and on the next page all calculations for one column of data and results.

Calculations and Results

Mass of water (g)					
ΔT of water (°C)					
ΔT of metal (°C)					
Heat flow (J) (absolute value)					
Specific heat (J/g · °C)					
Percentage error (copper only)					

Experiment 8

Report Sheet

Data

Run number					
Metal or unknown number					
Mass of metal (g)					
Temperature of boiling water (°C)					
Initial temperature of calorimeter water (°C)					
Final temperature (°C)					

Calculations and Results

Record all results in the table on the following page. Show below and on the next page all calculations for one column of data and results.

Calculations and Results

Mass of water (g)					
ΔT of water (°C)					
ΔT of metal (°C)					
Heat flow (J) (absolute value)					
Specific heat (J/g · °C)					
Percentage error (copper only)					

Chemical Names and Formulas: A Study Assignment

Performance Goal

9–1 Within the limits discussed in this exercise, and using a periodic table for reference, given the name (or formula) of any chemical species among the classifications below, write the formula (or name):

Elements in their stable form

Molecular binary compounds

Binary acids; oxyacids

Monatomic ions; polyatomic ions

Ionic compounds

INTRODUCTION

This study assignment presents a brief summary of the rules for writing formulas and naming substances commonly encountered in an introductory chemistry course. Basic definitions are stated, but theory relating to chemical bonding and the formation of ions is not considered. The purpose of this exercise is to practice writing formulas and names with help immediately available to clear up points that you may not understand. Hopefully you will *master* formula writing techniques during this laboratory period.

CHEMICAL OVERVIEW

Elements

This discussion will be limited to the more common elements listed in Figure 9.1. Given the name of one of these elements, you should be able to write its symbol, using a full periodic table for reference; given the symbol, you should be able to identify the element by name. This requires a certain amount of memorization, but the sheer memory work is reduced if you relate elemental names and symbols to the periodic table.

Figure 9.1

Partial periodic table showing the symbols and locations of the more common elements. The symbols above and the list that follows identify the elements you should be able to recognize or write, referring only to a complete periodic table. Associating the names and symbols with the table makes learning them much easier. The elemental names are:

aluminum	bismuth	cobalt	hydrogen	magnesium	oxygen	sodium
antimony	boron	chromium	iodine	manganese	phosphorus	strontium
argon	bromine	copper	iron	mercury	potassium	sulfur
arsenic	calcium	fluorine	krypton	neon	selenium	tin
barium	carbon	gold	lead	nickel	silicon	zinc
beryllium	chlorine	helium	lithium	nitrogen	silver	

Generally the chemical formula of an element in its stable form at room conditions is simply the symbol of the element. Seven gaseous elements, however, are not stable as individual atoms; two atoms combine to form a **diatomic molecule** as the unit particle of each of those elements. These elements and their correct chemical formulas are nitrogen, N_2; oxygen, O_2; hydrogen, H_2; fluorine, F_2; chlorine, Cl_2; bromine, Br_2; and iodine, I_2. Always remember to include the subscript 2 when writing the formulas of these substances *as elements, uncombined with any other elements.* In particular, notice that *this has nothing to do with these elements as they exist in compounds.* This is evident in the formulas for water, H_2O, and dinitrogen trioxide, N_2O_3.

Molecular Binary Compounds

Compounds made up of atoms held together entirely by covalent bonds are called **molecular compounds.** When a compound consists of two kinds of elements, it is called a **binary compound.** Molecular binary compounds therefore consist of two elements held together by covalent bonds. These elements are generally *both nonmetals.* You may use this identification

Table 9.1 *Prefixes Used in Naming Covalent Binary Compounds*

mono- = 1	hexa- = 6
di- = 2	hepta- = 7
tri- = 3	octa- = 8
tetra- = 4	nona- = 9
penta- = 5	deca- = 10

feature to distinguish molecular binary compounds from ionic binary compounds that will be discussed shortly.

Molecular binary compounds are identified by names consisting of two words. The main part of the first word is simply the name of the element appearing first in the formula; the main part of the second word is the name of the element appearing second in the formula, modified by an *-ide* suffix. The other part of each word is a prefix indicating the number of atoms of that particular element in the molecule. This is illustrated in the name dinitrogen trioxide for N_2O_3, in which *di-* is the prefix for 2 and *tri-* is the prefix for 3. A list of prefixes for numbers from 1 to 10 is given in Table 9.1. When the molecule contains only one atom of an element, the prefix *mono-* is frequently omitted, unless the species named is one of two or more compounds formed from the same two elements, such as CO, carbon monoxide, as compared to CO_2, carbon dioxide.

Acids

Inorganic acids, and some organic acids, are compounds that yield a hydrogen ion, or proton, when they ionize. (A proton and a hydrogen ion are the same thing. A hydrogen atom consists simply of a proton and an electron. When the electron is removed, producing a hydrogen ion, the only thing left is the proton.) Formulas of such acids are written with the ionizable hydrogen appearing first. This feature can usually be used to identify a formula as that of an acid.

A **binary acid** consists of hydrogen and one other nonmetallic element, usually in water solution. A binary acid is named by surrounding the root of the nonmetal with the prefix *hydro-* and the suffix *-ic*. Thus HCl is hydrochloric acid, the *chlor* coming from chlorine. The name *hydrosulfuric acid* suggests that the element other than hydrogen is sulfur. Its formula is H_2S.

Oxyacids contain oxygen as well as hydrogen and another nonmetal. The name of the most common oxyacid of each nonmetal is the root of the nonmetal followed by *-ic*. Thus H_2SO_4 is sulfuric acid, and the formula for the common oxyacid of chlorine, called chloric acid, is $HClO_3$. These names and formulas are somewhat similar to the names and formulas of the hydro-ic acids. Catch the distinction: *hydro-ic* acids have no oxygen, whereas *-ic* acids do contain oxygen.

There are six so-called *-ic* acids whose names and formulas you should memorize, because they constitute the base from which we will develop our approach to learning the names and formulas of a large number of chemical compounds. If you memorize these six acids, plus some prefixes and suffixes, you will be able to figure out all the other names and formulas without further memorization. The six acids are:

chloric, $HClO_3$	carbonic, H_2CO_3
sulfuric, H_2SO_4	phosphoric, H_3PO_4
nitric, HNO_3	acetic, $HC_2H_3O_2$

Acetic acid is the best known of a large group of organic acids that contain hydrogen but ionize only slightly in water. Organic chemists write the formulas for such acids differently, but for the purpose of this exercise we will follow the usual procedure of writing the ionizable hydrogen first.

The number of oxygen atoms may vary in oxyacids of the same nonmetal. Chlorine, for example, forms four oxyacids: $HClO_4$, $HClO_3$, $HClO_2$, and $HClO$. The names of these compounds are distinguished from each other by a series of prefixes and suffixes that are explained in Table 9.2. The key to the entire nomenclature system is the number of oxygen atoms *compared to the number in the -ic acid.* Study Table 9-2 to help you memorize these prefixes and suffixes and understand their use.

Nonmetals of the same chemical family frequently form acids that are similar in name and formula. Among the halogens, for example, HCl is hydrochloric acid, HF is hydrofluoric acid, HBr is hydrobromic acid, and HI is hydroiodic acid. The similarities extend to oxyacids for bromine and iodine, but not for fluorine, which forms no oxyacids. We thus find that $HBrO_2$ is bromous acid and HIO_4 is periodic acid.

Aside from the halogens, only sulfur and nitrogen form important oxyacids other than their well-known *-ic* acids. In both cases it is the *-ous* acid that is formed, each with one less oxygen atom than is present in the *-ic* acid. Thus HNO_2 is the formula for nitrous acid, and sulfurous acid has the formula H_2SO_3. Selenium and tellurium, atomic numbers 34 and 52, in the same column of the periodic table as sulfur, form corresponding *-ous* acids.

Table 9.2 *Names of Oxyacids and Oxyanions of Chlorine (HCl included for comparison)*

I	*II*	*III*	*IV*	*V*	*VI*	*VII*
Acid Name	*Acid Suffixes and Prefixes*	*Acid Formula*	*Oxygens Compared to -ic Acid*	*Ion Name*	*Ion Suffixes and Prefixes*	*Ion Formula*
hydrochloric (binary acid)	*hydro-ic*	HCl	no oxygen	chloride	*-ide* named as monatomic anion	Cl^-
hypochlorous	*hypo-ous*	HClO	-2	hypochlorite	*hypo-ite*	ClO^-
chlorous	*-ous*	$HClO_2$	-1	chlorite	*-ite*	ClO_2^-
chloric	*-ic*	$HClO_3$	same	chlorate	*-ate*	ClO_3^-
perchloric	*per-ic*	$HClO_4$	$+1$	perchlorate	*per-ate*	ClO_4^-

Oxidation State: Oxidation Number

Chemists use a set of **oxidation numbers,** or consider the **oxidation state** of an element, in discussing oxidation–reduction reactions. These numbers are also part of the modern nomenclature system. The rules by which these numbers are assigned are as follows:

1. The oxidation number of any elemental substance is zero.

2. The oxidation number of a monatomic ion is the same as the charge on the ion.

3. The oxidation number of combined oxygen is −2, except in peroxides (−1) and superoxides ($-\frac{1}{2}$).

    We will not encounter peroxides or superoxides in this assignment.

4. The oxidation number of combined hydrogen is +1, except in hydrides (−1).

5. In any molecular or ionic species, the sum of the oxidation numbers of all atoms in the species is equal to the charge on the species.

6. In a compound, the oxidation numbers of all atoms add up to zero.

The manner in which these rules are applied will be discussed as the need arises.

Monatomic Ions

A monatomic ion is a single atom that has acquired an electrical charge by gaining or losing one, two, or three electrons. Its formula is the symbol of the element followed by a superscript indicating the charge. For example, the formula of a calcium ion is Ca^{2+}, and for a chloride ion, Cl^-. It is important that the charge be indicated for an ion. Without that charge, the formula would be that of an electrically neutral atom from which the ion was formed, a very different species with very different chemical properties. Ions with a negative charge are called **anions;** ions with a positive charge are called **cations.**

The nonmetals in Groups VA, VIA, and VIIA form monatomic anions by gaining electrons. Ions from Group VA elements have a 3− charge, as in N^{3-}; from Group VIA, a 2− charge, as in O^{2-}; and from Group VIIA, a 1− charge, as in F^-. The name of a monatomic anion is simply the name of the element, modified by an *-ide* suffix, as in nitride, oxide, or fluoride.

Metals in Groups IA, IIA, and IIIA form cations with charges of 1+, 2+, and 3+, respectively. Many metals in the B groups of the periodic table form two monatomic ions that differ in charge. The best example is iron, which yields the Fe^{2+} and Fe^{3+} ions. These ions are distinguished by adding the oxidation state, or charge, to the name of the element. Accordingly, Fe^{2+} is the iron(II) ion, and Fe^{3+} is the iron(III) ion. Notice how these names are written; the oxidation state is written in Roman numerals *and enclosed in parentheses* immediately after the name of the element, with no space between the name and the parentheses.

> **CAUTION**
>
> **Students often neglect to enclose the oxidation state in parentheses; the name is not correctly written if the parentheses are missing.**

The names of iron(II) and iron(III) ions are pronounced "iron two" and "iron three," respectively.

1+	2+											3+		3−	2−	1−
H^+																H^-
Li^+	Be^{2+}													N^{3-}	O^{2-}	F^-
Na^+	Mg^{2+}											Al^{3+}		P^{3-}	S^{2-}	Cl^-
K^+	Ca^{2+}				Cr^{2+} Cr^{3+}	Mn^{2+} Mn^{3+}	Fe^{2+} Fe^{3+}	Co^{2+} Co^{3+}	Ni^{2+}	Cu^+ Cu^{2+}	Zn^{2+}			As^{3-}	Se^{2-}	Br^-
	Sr^{2+}									Ag^+			Sn^{2+} Sn^{4+}	Sb^{3+}		I^-
	Ba^{2+}										Hg_2^{2+} Hg^{2+}		Pb^{2+} Pb^{4+}	Bi^{3+}		

NH_4^+

Figure 9.2

Partial periodic table of common ions

Notes: (1) Tin (Sn) and lead (Pb) form monatomic ions in a +2 oxidation state. In their +4 oxidation states they are more accurately described as being covalently bonded but such compounds are frequently named as if they were ionic compounds; (2) Hg_2^{2+} is a diatomic elemental ion. Its name is mercury(I), indicating a +1 charge from each atom in the diatomic ion; (3) ammonium ion, NH_4^+, is included as the only other common polyatomic cation, thereby completing this table as a minimum list of the cations you should be able to recall simply by referring to a full periodic table.

Notice that oxidation states in the names of monatomic ions are used only to distinguish between ions of the same element that have different charges. Oxidation numbers are not used if a metal forms only one kind of ion.

The cations formed by mercury require special comment. The mercury(II) ion, Hg^{2+}, is a typical monatomic ion. There is also a mercury(I) ion, but it is diatomic. Its formula is Hg_2^{2+}. The mercury(I) name is logical if you realize that *each atom* is contributing a 1+ charge to the diatomic ion.

Figure 9.2 locates in a periodic table the monatomic ions you should be able to recognize on sight, or write if given the name of the ion.

Polyatomic Anions Derived from the Total Ionization of Oxyacids

When an oxyacid ionizes, the resulting anion has more than one atom; it is a *polyatomic anion*. These are *oxyanions,* so called because they contain oxygen. Names of oxyanions are related to the acid from which they come; the prefix or suffix of the acid is replaced by a prefix or suffix for the anion. The system is illustrated for chlorine in Table 9.2, page 110. Memorize these prefixes and suffixes, and you will be able to apply them to a large number of compounds, including many you may never have heard of before.

The negative charge on an ion from the total ionization of an oxyacid is equal to the number of hydrogen atoms in the neutral acid molecule. Chloric acid, with one hydrogen, produces an oxyanion with a single negative charge, ClO_3^-; sulfuric acid, with two hydrogens, yields the double negative sulfate ion, SO_4^{2-}; and removal of three hydrogens from phosphoric acids yields an ion with a 3− charge, PO_4^{3-}.

Oxyanions Derived from the Stepwise Ionization of Polyprotic Acids

When an acid containing two or more hydrogen atoms ionizes, it loses the hydrogen ions one by one. There are, therefore, intermediate ions that contain hydrogen. The stepwise ionization of sulfuric acid may be represented by

$$H_2SO_4 \xrightarrow{-H^+} HSO_4^- \xrightarrow{-H^+} SO_4^{2-}$$

The HSO_4^- ion can be thought of as a sulfate ion with a hydrogen attached. It is given the logical name *hydrogen sulfate ion*. When triprotic phosphoric acid, H_3PO_4, ionizes, there are two intermediate ions, $H_2PO_4^-$ and HPO_4^{2-}. The first of these is the phosphate ion with two hydrogens attached, so it is called the dihydrogen phosphate ion, which distinguishes it from HPO_4^{2-}, the hydrogen (or monohydrogen) phosphate ion. Intermediate ions from other polyprotic acids are named in a similar manner.

Other Polyatomic Ions

There are two other polyatomic ions that are so common you should recognize them instantly. These are the ammonium ion, NH_4^+, and the hydroxide ion, OH^-. Many other polyatomic ions exist, but it is not necessary that they be memorized at this time unless your instructor directs you to do so. Some of them are listed in Tables 9.3 and 9.4, which include most of the ions you are apt to encounter in a beginning chemistry course.

Ionic Compounds

Two rules govern the nomenclature of ionic compounds:

1. The name of an ionic compound is the name of the positive ion followed by the name of the negative ion.

2. The formula of an ionic compound is the formula of the positive ion followed by the formula of the negative ion, each taken as many times as may be necessary to bring the total charge to zero.

To name an ionic compound when given the formula, you need only to recognize the ions present. You must be familiar with the number of oxygen atoms in the various oxyanions, as well as the rules by which the anions are named. For a compound having a cation from a metal that forms two different monatomic ions, you must apply the oxidation-state rules to determine which of those ions is present. If the compound is $FeCl_2$, for example, you must recognize that the chloride ion has a 1− charge. There are two chloride ions present, so that the total negative charge in the formula unit is 2−. The sum of all oxidation numbers in the formula must be zero, which means the 2+ charge must come from the iron ion, and the compound must therefore be iron(II) chloride. Similar reasoning would lead to the conclusion that $FeCl_3$ is iron(III) chloride.

Table 9.3 *Common Cations*

Ionic Charge: +1	Ionic Charge: +2	Ionic Charge: +3
Alkali Metals: *Group IA*	*Alkali Earths:* *Group IIA*	*Group IIIA*
Li^+ — Lithium	Be^{2+} — Beryllium	Al^{3+} — Aluminum
Na^+ — Sodium	Mg^{2+} — Magnesium	Bi^{3+} — Bismuth
K^+ — Potassium	Ca^{2+} — Calcium	Sb^{3+} — Antimony
Rb^+ — Rubidium	Sr^{2+} — Strontium	*Transition Elements*
Cs^+ — Cesium	Ba^{2+} — Barium	Cr^{3+} — Chromium(III)
Transition Elements	*Transition Elements*	Fe^{3+} — Iron(III)
Cu^+ — Copper(I)	Cr^{2+} — Chromium(II)	Co^{3+} — Cobalt(III)
Ag^+ — Silver	Mn^{2+} — Manganese(II)	
Polyatomic Ions	Fe^{2+} — Iron(II)	
NH_4^+ — Ammonium	Co^{2+} — Cobalt(II)	
Others	Ni^{2+} — Nickel	
H^+ — Hydrogen	Cu^{2+} — Copper(II)	
or	Zn^{2+} — Zinc	
H_3O^+ — Hydronium	Cd^{2+} — Cadmium	
	Hg_2^{2+} — Mercury(I)	
	Hg^{2+} — Mercury(II)	
	Others	
	Sn^{2+} — Tin(II)	
	Pb^{2+} — Lead(II)	

In writing the formulas of compounds in which a polyatomic ion appears more than once, the entire ion is enclosed in parentheses, followed by a subscript indicating the number of ions in the formula unit. For example, the formula of calcium nitrate is $Ca(NO_3)_2$. This is the only time parentheses are used. Specifically, they are not used when a polyatomic ion appears only once in the formula, as in calcium sulfate, $CaSO_4$. Nor is the symbol of a monatomic ion enclosed in parentheses just because it happens to have two letters, as in calcium bromide, $CaBr_2$.

Table 9.4 *Common Anions*

Ionic Charge: −1		Ionic Charge: −2	Ionic Charge: −3
Halogens: Group VIIA	*Oxyanions*	*Group VIA*	*Group VA*
F^- Fluoride	ClO_4^- Perchlorate	O^{2-} Oxide	N^{3-} Nitride
Cl^- Chloride	ClO_3^- Chlorate	S^{2-} Sulfide	P^{3-} Phosphide
Br^- Bromide	ClO_2^- Chlorite	*Oxyanions*	*Oxyanion*
I^- Iodide	ClO^- Hypochlorite	CO_3^{2-} Carbonate	PO_4^{3-} Phosphate
Acidic Anions		SO_4^{2-} Sulfate	PO_3^{3-} Phosphite
HCO_3^- Hydrogen carbonate	BrO_3^- Bromate	SO_3^{2-} Sulfite	
	BrO_2^- Bromite	$C_2O_4^{2-}$ Oxalate	
HS^- Hydrogen sulfide	BrO^- Hypobromite	CrO_4^{2-} Chromate	
HSO_4^- Hydrogen sulfate	IO_4^- Periodate	$Cr_2O_7^{2-}$ Dichromate	
	IO_3^- Iodate		
	IO^- Hypoiodite	*Acidic Anion*	
HSO_3^- Hydrogen sulfite	NO_3^- Nitrate	HPO_4^{2-} Monohydrogen phosphate	
	NO_2^- Nitrite		
$H_2PO_4^-$ Dihydrogen phosphate	OH^- Hydroxide	*Diatomic*	
	$C_2H_3O_2^-$ Acetate	O_2^{2-} Peroxide	
Other Anions	MnO_4^- Permanganate		
SCN^- Thiocyanate			
CN^- Cyanide			
H^- Hydride			

Experiment 9

Advance Study Assignment

1. Give the names and formulas of two oxyacids of chlorine.

2. Give the formula and name of an ionic compound containing (a) the sulfate ion, (b) the nitrite ion.

3. Write the formula of an ionic compound (using the symbols X and Y) derived from:

   a. X^+ cation and Y^{3-} anion.

   b. X^{2+} cation and HY^- anion.

   c. X^{3+} cation and HY^{2-} anion.

Experiment 9

Work Page

General Instructions: For each substance whose name is given, write the formula; if the formula is given, write the name. Unless stated otherwise, a periodic table should be your only reference.

Elements

Write the formulas of the elements in their natural, stable states.

Iron	Na
Calcium	Cl_2
Nitrogen	Cu
Bromine	Mg
Potassium	Ni

Molecular Binary Compounds

Carbon dioxide	CBr_4
Dinitrogen tetroxide	CO
Iodine chloride	P_2O_3
Sulfur trioxide	SiS_2
Diphosphorus pentoxide	S_2F_6

Acids

Hydrobromic acid	HClO
Sulfuric acid	HI
Bromic acid	HNO_3
Phosphoric acid	H_2SO_3
Nitrous acid	HIO_4
Perchloric acid	$HBrO_2$

Monatomic and Polyatomic Ions

Calcium ion	Fe^{2+}
Sulfate ion	Br^-
Monohydrogen phosphate ion	ClO^-
Nitrite ion	$CO_3{}^{2-}$
Iron(III) ion	Cr^{3+}
Iodite ion	$SO_3{}^{2-}$
Sulfide ion	$HCO_3{}^-$

Experiment 9

Work Page

Ionic Compounds

Sodium nitrate	K_2SO_4
Calcium fluoride	Na_3PO_4
Potassium hydrogen sulfate	$Pb(NO_3)_2$
Sodium carbonate	$FeCl_3$
Potassium bromide	KIO_3
Iron(III) sulfide	$Ca(OH)_2$
Magnesium chloride	$Al_2(SO_4)_3$
Sodium dihydrogen phosphate	$HgCO_3$
Ammonium sulfate	$NaClO_2$
Copper(II) carbonate	KHS
Barium hydroxide	K_2O
Silver bromide	$NaHSO_3$
Mercury(II) sulfate	$(NH_4)_2CO_3$
Potassium nitrite	FeO
Calcium chlorate	$NaHCO_3$
Iron(II) hydroxide	CaI_2
Copper(I) iodate	NH_4Br
Aluminum sulfite	$BaCl_2$
Magnesium oxide	$FePO_4$
Lead(II) iodide	Ag_2SO_4
Sodium hypochlorite	$Co(OH)_2$

Ionic Compounds (Continued)

Lithium hydrogen sulfite	NH_4NO_2
Ammonium carbonate	Cu_2O
Mercury(I) chloride	K_3PO_4
Aluminum oxide	$(NH_4)_2HPO_4$
Potassium periodate	$AgBrO_3$

Experiment 9

Work Sheet

Compounds Containing Less Common Ions

(Refer to tables of cations and anions and the periodic table when writing these formulas.)

Strontium sulfate (strontium, atomic number 38)	
Cesium iodide (cesium, atomic number 55)	
Indium chloride (indium, atomic number 49)	
Tellurium trioxide (tellurium, atomic number 52)	
Calcium hydride	
Sodium cyanide	
Iron(III) thiocyanate	
Nickel(II) chromate	

Experiment 9

Report Sheet

General Instructions: For each substance whose name is given, write the formula; if the formula is given, write the name. Unless stated otherwise, a periodic table should be your only reference.

Elements

Write the formulas of the elements in their natural, stable states.

Iron	Na
Calcium	Cl_2
Nitrogen	Cu
Bromine	Mg
Potassium	Ni

Molecular Binary Compounds

Carbon dioxide	CBr_4
Dinitrogen tetroxide	CO
Iodine chloride	P_2O_3
Sulfur trioxide	SiS_2
Diphosphorus pentoxide	S_2F_6

Acids

Hydrobromic acid	HClO
Sulfuric acid	HI
Bromic acid	HNO_3
Phosphoric acid	H_2SO_3
Nitrous acid	HIO_4
Perchloric acid	$HBrO_2$

Monatomic and Polyatomic Ions

Calcium ion	Fe^{2+}
Sulfate ion	Br^-
Monohydrogen phosphate ion	ClO^-
Nitrite ion	CO_3^{2-}
Iron(III) ion	Cr^{3+}
Iodite ion	SO_3^{2-}
Sulfide ion	HCO_3^-

Experiment 9

Report Sheet

Ionic Compounds

Sodium nitrate	K_2SO_4
Calcium fluoride	Na_3PO_4
Potassium hydrogen sulfate	$Pb(NO_3)_2$
Sodium carbonate	$FeCl_3$
Potassium bromide	KIO_3
Iron(III) sulfide	$Ca(OH)_2$
Magnesium chloride	$Al_2(SO_4)_3$
Sodium dihydrogen phosphate	$HgCO_3$
Ammonium sulfate	$NaClO_2$
Copper(II) carbonate	KHS
Barium hydroxide	K_2O
Silver bromide	$NaHSO_3$
Mercury(II) sulfate	$(NH_4)_2CO_3$
Potassium nitrite	FeO
Calcium chlorate	$NaHCO_3$
Iron(II) hydroxide	CaI_2
Copper(I) iodate	NH_4Br
Aluminum sulfite	$BaCl_2$
Magnesium oxide	$FePO_4$
Lead(II) iodide	Ag_2SO_4
Sodium hypochlorite	$Co(OH)_2$

Ionic Compounds (Continued)

Lithium hydrogen sulfite	NH_4NO_2
Ammonium carbonate	Cu_2O
Mercury(I) chloride	K_3PO_4
Aluminum oxide	$(NH_4)_2HPO_4$
Potassium periodate	$AgBrO_3$

Experiment 9

Report Sheet

Compounds Containing Less Common Ions

(Refer to tables of cations and anions and the periodic table when writing these formulas.)

Strontium sulfate (strontium, atomic number 38)	
Cesium iodide (cesium, atomic number 55)	
Indium chloride (indium, atomic number 49)	
Tellurium trioxide (tellurium, atomic number 52)	
Calcium hydride	
Sodium cyanide	
Iron(III) thiocyanate	
Nickel(II) chromate	

Chemical Equations: A Study Assignment

Performance Goal

10–1 Given information from which you can write formulas for all reactants and all products for each of the following types of reactions, write the balanced chemical equation for the reactions:

Double replacement

Double replacement—precipitation

Double replacement—acid/base (neutralization)

Combination (synthesis)

Decomposition

Complete oxidation or burning of organic compounds (combustion)

Single replacement (redox)

Other reactions in which reactants and products are identified

CHEMICAL OVERVIEW

A chemist uses a chemical equation to describe a chemical change. The general form of a chemical equation is

$$\text{Reactant 1} + \text{Reactant 2} + \ldots \rightarrow \text{Product 1} + \text{Product 2} + \ldots$$

The substances that enter into the reaction are called **reactants**. They are identified by their chemical formulas, written on the left side of the equation, and separated from each other by plus signs. The formulas of the new substances produced in the reaction, called **products**, are written on the right side, again separated by plus signs. The two sides of the equation are separated by an arrow pointing from the reactants to the products, indicating that the reactants are changed into the products. In reading a chemical equation, or expressing it in words, the arrow is frequently read as "yields," "produces," or "forms"; any other term that suggests the creation of a substance not originally present is equally satisfactory.

Symbols are frequently added to chemical equations to indicate the conditions under which the reaction occurs. The symbols (s), (*l*), or (g) immediately after the formula of a substance indicate that the substance is in the solid, liquid, or gaseous state, respectively. A substance that is in aqueous (water) solution may have (aq) after its formula. Sometimes the arrow between reactants and products is lengthened, and words, formulas, temperatures, or other symbols are written above (or above and below) the arrow to indicate reaction conditions or other substances in the reaction vessel. None of these supplementary items will be used in this exercise; however, if your instructor requests that you use them, you should, of course, follow his or her directions.

A chemical equation does two things. First, it tells you what substances are involved in a chemical change. To do this accurately, it is essential that the substances be represented by their correct chemical formulas. It is assumed in this exercise that, given the name of a chemical, you are able to write its formula. Second, an equation has quantitative significance. It obeys the law of conservation of mass, which indicates that the total mass of all the reactants is equal to the total mass of all the products in an ordinary chemical change. In order for this to be true, the equation must have equal numbers of atoms of each individual element on the two sides of the equation. The equation is then said to be **balanced**.

These two characteristics of an equation lead to a simple two-step procedure by which an equation may be written:

1. Write the correct chemical formula for each reactant on the left and each product on the right.

2. *Using coefficients only*, balance the number of atoms of each element on each side of the equation. If no coefficient is written, its value is assumed to be one (1).

It is impossible to overemphasize the importance of following these two steps literally, and keeping them independent. In Step 1, write the correct formulas without concern about where the atoms come from, or how many atoms of an element may be present in some species on the other side of the equation. In Step 2, be sure that you balance the atoms of each element by placing whole-number coefficients in front of chemical formulas, and by no other means. Specifically,

DO NOT   change a correct chemical formula in order to balance an element;

DO NOT   add some real or imaginary chemical species to either side to make an element balance.

Quite often the word description of a chemical reaction will not identify all of the species that must be included in the equation. If you are familiar with the kinds of reactions described in the performance goal, you will be able to identify the substances not mentioned. The reaction types will be discussed as they are encountered.

EXAMPLES

The following examples are in the form of a program in which you learn by answering a series of questions. Obtain an opaque shield (a piece of cardboard, or a folded piece of paper you cannot see through) that is wide enough to cover this page. In each example place the shield on the book page so it covers everything beneath the first dotted line that runs across the page. Read to that point, and write in the space provided whatever is asked. Then lower the shield to the next dotted line. The material exposed will begin with the correct response to the question you have just answered. Compare this answer to yours, looking back to correct any misunderstanding if the two are different. When you fully understand the first step, read to the next dotted line and proceed as before.

A. Combination Reactions

A combination reaction occurs when two or more substances combine to form a single product. The reactants may be elements or compounds, perhaps one or more of each. Quite often the description of the reaction will give the chemical name of the product only. For example, the equation for the reaction in which sodium chloride is formed from its elements is $2\ Na + Cl_2 \rightarrow 2\ NaCl$. An example of a combination reaction between compounds is $CaO + H_2O \rightarrow Ca(OH)_2$.

Example 1

Write the equation showing how magnesium oxide is formed from its elements.

"magnesium oxide is formed" indicates that magnesium oxide is the product of the reaction, so its formula will appear on the right side of the equation. "from its elements" identifies magnesium and oxygen as the reactants whose formulas will be written to the left of the arrow. Complete Step 1 of the procedure by writing the unbalanced equation.

....

1a. $Mg + O_2 \rightarrow MgO$

Remember, oxygen is a diatomic element; its correct formula is therefore O_2 and not simply O.

Step 2 calls for you to balance the atoms of each element on the two sides of the equation. As it stands, there is one magnesium atom on each side; magnesium is in balance. The left side of the equation has two atoms of oxygen, and the right side only one. What must you do to balance oxygen? Remember, there is only one way to do it, and watch out for the DON'Ts listed earlier. Balance the oxygen.

$$Mg + O_2 \rightarrow MgO$$

....

1b. $Mg + O_2 \rightarrow 2\ MgO$

Remember, your only way to balance atoms of an element in Step 2 is to use coefficients in front of substances in the unbalanced equation. Some common WRONG responses to the above—and what is wrong with them—are:

$Mg + O_2 \rightarrow MgO + O$ There are no oxygen atoms in the reaction. This violates the second DON'T.

$$Mg + O_2 \rightarrow MgO_2$$ MgO_2 happens to be a real substance, but it is *not* the product of this reaction. This violates the first DON'T.

$$Mg + O_2 \rightarrow Mg\ 2\ O$$ Mg 2 O is not a chemical formula. Coefficients are placed in front of a formula, not in the middle, and they affect the entire formula.

The last of the three wrong balancing methods points out that in balancing oxygen we have *un*balanced magnesium. There is now one magnesium atom on the left and two on the right. Correct this now.

....

1c. $2\,Mg + O_2 \rightarrow 2\,MgO$

There is another way you might have balanced $Mg + O_2 \rightarrow MgO$. You could have introduced the fractional coefficient 1/2 in front of oxygen: $Mg + 1/2\,O_2 \rightarrow MgO$. Fractional coefficients should not be used in this exercise. They may be used as a means to the final equation, however. If you do choose to balance the equation with a fractional coefficient, as above, you can then multiply the entire equation by 2 (doubling *each* coefficient), giving $2\,Mg + O_2 \rightarrow 2\,MgO$. Incidentally, equations should be written with the *smallest* whole-number coefficients. If, in your balancing procedure, you happened to arrive at $4\,Mg + 2\,O_2 \rightarrow 4\,MgO$, you could divide the entire equation—each coefficient—by 2 to get the desired result.

Example 2

Write the equation for the formation of iron(III) oxide from its elements.

Complete the first step by writing the formulas of the reactants on the left and the formula of the product on the right.

....

2a. $Fe + O_2 \rightarrow Fe_2O_3$

Start with the iron; balance it first and leave oxygen unbalanced.

....

2b. $2\,Fe + O_2 \rightarrow Fe_2O_3$

There are two thought processes by which balancing may be completed, both leading to the same result. Both will be discussed—after you have balanced the rest of the equation yourself.

....

2c. $4\,Fe + 3\,O_2 \rightarrow 2\,Fe_2O_3$

Oxygen atoms come two to the package in O_2 molecules, and three to the package in Fe_2O_3 units. Six atoms—2 times 3—is the smallest number of atoms by which a 3-and-2 combination can be equalized. If you take 3 packages of 2 each, you will have the same number as 2 packages of 3 each. This fixes the coefficients of O_2 and Fe_2O_3. The coefficient of iron is adjusted to correspond with the iron atoms in 2 Fe_2O_3.

A second way of reaching the final equation is to select the fractional coefficient of O_2 that will give the proper number of oxygen atoms to balance the three on the right side of $2\,Fe + O_2 \rightarrow Fe_2O_3$. With three oxygens on the right, we need three on the left, where they come two to a package in O_2. We therefore need $1\frac{1}{2}$ packages, or 3/2, yielding $2\,Fe + 3/2\,O_2 \rightarrow Fe_2O_3$. This balanced equation can be cleared of fractions by multiplying all coefficients by 2, giving $4\,Fe + 3\,O_2 \rightarrow 2\,Fe_2O_3$.

It is worthwhile to become familiar with both methods. The 3-and-2 combination appears frequently enough to justify the routine 2-of-3 and 3-of-2 thought processes. It is convenient to realize that if you need X atoms of oxygen from O_2 molecules, the number of molecules required is $X/2$. Doubling the equation yields whole-number coefficients.

B. Decomposition Reactions

The chemical change in which a single reactant decomposes into two or more products is a decomposition reaction. This is just the opposite of a combination reaction; indeed, many combination reactions can be reversed, as $2\,NaCl \rightarrow 2\,Na + Cl_2$. The reaction $2\,Al(OH)_3 \rightarrow Al_2O_3 + 3\,H_2O$ illustrates a decomposition of a compound into two simpler compounds. Another type of decomposition reaction occurs when hydrates (compounds containing water of hydration) are heated. $Na_2CO_3 \cdot 10\,H_2O \xrightarrow{\Delta} Na_2CO_3 + 10\,H_2O$ illustrates such a reaction, where Δ written over the arrow generally means "applying heat."

Example 3

Calcium carbonate is decomposed into calcium oxide and carbon dioxide by heat. Write the equation.

The first step is to write the formulas of reactants and products in their proper places. Proceed that far.

....

3a. $CaCO_3 \rightarrow CaO + CO_2$

Now Step 2: balance the atoms of each element on the two sides of the equation.

....

3b. $CaCO_3 \rightarrow CaO + CO_2$

Sometimes balancing an equation is easy—particularly when all coefficients are 1!

C. Complete Oxidation or Burning of Organic Compounds

Other than the oxides of carbon, carbonates, and a few other substances, the compounds of carbon are classified as **organic compounds**. Hydrogen is almost always present in an organic compound, and oxygen is a third very common element. When compounds containing carbon and hydrogen, or carbon, hydrogen, and oxygen, react *completely* with an excess of oxygen, the products are always carbon dioxide and water. Such a reaction may occur with the oxygen in the air, giving heat and light, in which case the process is called **burning**; and it may occur in living organisms, again giving off heat and other forms of energy, in which case it is referred to as **oxidation**. The description of such a reaction may be very brief: Compound *X* is completely oxidized, or Compound *Y* is burned in air. In both reactions you must recognize oxygen as an unnamed reactant to be included in the equation, and write the formulas of carbon dioxide and water as the products. $2\,C_6H_{14} + 19\,O_2 \rightarrow 12\,CO_2 + 14\,H_2O$ is an example of a burning reaction. Because organic compounds are frequently quite large, equations may have large coefficients; but don't let that bother you, because they are reached by the same method outlined above.

Example 4

Write the equation for the complete oxidation of methyl ethyl ketone, $CH_3COC_2H_5$.

Methyl ethyl ketone has been chosen for this example because its equation includes all the little things you must look out for in writing oxidation equations. First, notice that organic chemists sometimes write formulas in ways that seem strange to the beginning student. This is because the sequences of elements and certain combinations in the formula suggest how atoms are arranged in the molecule and identify the kind of compound it is. Second, you must be sure to count *all* the atoms of a given element in a molecular formula when balancing, such as 4 carbon atoms, 8 hydrogen atoms, and 1 oxygen atom in a molecule of $CH_3COC_2H_5$. A third point will show up later. Right now, complete Step 1 by writing the formulas of reactants and products in their proper places in an unbalanced equation.

....

4a. $CH_3COC_2H_5 + O_2 \rightarrow CO_2 + H_2O$

Always remember that, although it is unnamed in the statement of the reaction, oxygen is a second reactant and the products are carbon dioxide and water.

To begin Step 2, you balance both carbon and hydrogen. With the warning already given, add those coefficients to the equation.

....

4b. $CH_3COC_2H_5 + O_2 \rightarrow 4CO_2 + 4H_2O$

With carbon and hydrogen balanced, and oxygen in its elemental form on the left, oxygen can be balanced simply by placing in front of oxygen the coefficient that does the job. Sounds simple, but be careful . . .

....

4c. $2\ CH_3COC_2H_5 + 11\ O_2 \rightarrow 8\ CO_2 + 8\ H_2O$

Starting from $CH_3COC_2H_5 + O_2 \rightarrow 4\ CO_2 + 4\ H_2O$, you count 12 oxygen atoms on the right side of the equation. On the left, *one of the required 12 oxygen atoms comes from the reactant,* and the remaining 11 come from O_2. This is the third thing you must look out for in balancing oxidation equations, being sure not to overlook oxygen present in the compound being oxidized. With 11 oxygen atoms to come from O_2, you can balance the equation with a fractional coefficient: $CH_3COC_2H_5 + 11/2\ O_2 \rightarrow 4\ CO_2 + 4\ H_2O$. Doubling the entire equation gives whole-number coefficients, as required.

D. Oxidation–Reduction Reactions

The words **oxidize** and **oxidation** have meaning in chemistry other than "reaction with oxygen," as suggested in the foregoing section. In its broader meaning, oxidation means loss of electrons. If one reactant loses electrons, another reactant must gain those electrons. The process of gaining electrons is called **reduction**. A reaction in which oxidation and reduction occur—and they must always occur simultaneously—is called an **oxidation–reduction reaction**, frequently shortened to "redox" reaction.

In this exercise we will be concerned with only one kind of redox reaction. The equation has the appearance of an element reacting with a compound in such a manner that the element replaces one of the elements in the compound. $Zn + Cu(NO_3)_2 \rightarrow Cu + Zn(NO_3)_2$ is such a reaction. It appears as if elemental zinc has replaced copper from $Cu(NO_3)_2$. This kind of equation is frequently called a **single replacement equation**. Given an element and an ionic compound as reactants, you should recognize the possibility of a redox reaction and be able to write the single replacement equation for that reaction. Whether or not the reaction actually occurs requires laboratory confirmation, of course.

Example 5

Gaseous hydrogen is released when zinc reacts with hydrochloric acid. Write the equation for the reaction.

The reactants and one of the products are identified. As you write the unbalanced equation (Step 1) for these three species, see if you can recognize the single replacement character of that equation and then figure out the formula of the second product.

....

5a. $Zn + HCl \rightarrow H_2 + ZnCl_2$

In the reaction zinc is releasing, or replacing, hydrogen in HCl. The second product is therefore zinc chloride. Balancing the equation is straightforward . . .

....

5b. $Zn + 2\ HCl \rightarrow H_2 + ZnCl_2$

Example 6

Write the equation for the reaction between aluminum and nickel nitrate.

This time you are given only the names of two reactants. Write their formulas on the left side of the arrow, leaving the product side blank.

....

6a. Al + Ni(NO$_3$)$_2$ →

Here's where your skill in recognizing the possibility of a redox reaction comes into play. What possible products could come from these reactants? There is no indication that the nitrate ion decomposes. Ions of aluminum and nickel are both positively charged, so there is no way they could form a compound. The nitrate ion is negatively charged, so it could form a compound with an aluminum ion. It all points to a single replacement equation in which aluminum bumps nickel out of the compound. Complete Step 1 by writing the formulas of the products on the right side of the equation.

....

6b. Al + Ni(NO$_3$)$_2$ → Ni + Al(NO$_3$)$_3$

This example gives us an opportunity to introduce an important technique in balancing equations. A quick glance shows that aluminum and nickel are balanced, but nitrogen and oxygen are not. You could balance them individually, but there is an easier way. It was noted above that the *nitrate ion* does not decompose in the reaction; in other words, the nitrate ion is the same on the product side of the equation as it is on the reactant side. Any time a polyatomic (many atom) ion is unchanged in a chemical reaction, that ion may be balanced *as a unit* in the equation. In other words, your thought process should be, "There are two nitrate ions on the left, and three nitrate ions on the right. How do I balance them?" How *do* you balance a 3-and-2 combination? You already know that, so go ahead. While you're at it, be sure to do whatever is necessary to keep the aluminum and nickel in balance.

....

6c. 2 Al + 3 Ni(NO$_3$)$_2$ → 3 Ni + 2 Al(NO$_3$)$_3$

You need 3 nickel nitrate units, where nitrate ions appear in packages of two, to balance 2 aluminum nitrate units, where the nitrate ions appear in packages of three, giving 6 nitrate ions on each side of the equation. The coefficients for the metals complete the equation.

E. Double Replacement Reactions

As the name suggests, double replacement reactions involve the combination of ions from different sources to form a new product. If two ions are to combine, one must have a positive charge and the other must have a negative charge. The combination of a lead ion from lead(II) nitrate and a chloride ion from sodium chloride to form lead(II) chloride is a good example: Pb(NO$_3$)$_2$ + 2 NaCl → PbCl$_2$ + 2 NaNO$_3$. If you look at the equation, it appears as if the positive and negative ions in the two reactants have simply "changed partners" in the products; the positive ion of the first reactant has joined up with the negative ion of the second reactant, and the negative ion of the first has combined with the positive ion of the second. Whenever you see an equation with two ionized reactants, you can make an intelligent prediction that the products will be derived from an exchange of ions, and write their formulas accordingly.

Most double replacement reactions occur in water solution. One of the driving forces for these reactions is the formation of an insoluble ionic solid, called a **precipitate**. In the example above, lead(II) chloride is insoluble in water, so it precipitates as the ions combine with each other.

The other driving force that brings ions together is the formation of a molecular product, in which a covalent (shared electron pair) bond forms between the reacting ions. The most common molecular product is water, as in $HCl + KOH \rightarrow KCl + H_2O$. This kind of reaction, in which an acid reacts with a base, is called a **neutralization reaction**. The ionic product formed (KCl, in this example) is classified as a **salt**.

Experiment 10

Work Page

Equation-Writing Exercise

Write the chemical equation for each reaction described below.

A. Combination Reactions

1. Diphosphorus trioxide is formed by direct combination of its elements.

2. Ammonia and sulfuric acid combine to form ammonium sulfate.

B. Decomposition Reactions

3. Ammonium nitrite decomposes into nitrogen and water.

4. When heated, potassium chlorate decomposes into oxygen and potassium chloride.

C. Complete Oxidation or Burning of Organic Compounds

5. Propane, C_3H_8, burns in air.

6. Acetaldehyde, CH_3CHO, is completely oxidized.

D. Oxidation–Reduction Reactions

7. Hydrogen is released when aluminum reacts with hydrochloric acid.

8. Magnesium reacts with silver nitrate solution.

E. Double Replacement Reactions

9. Barium carbonate precipitates from the reaction of barium chloride and sodium carbonate solutions.

10. Sulfuric acid reacts with calcium hydroxide.

11. Sodium iodate and silver nitrate solutions are combined.

12. Potassium fluoride reacts with hydrobromic acid.

13. Zinc hydroxide reacts with hydrochloric acid.

Experiment 10

Work Page

F. Other Reactions

14. Copper(II) chloride and water result from the reaction of copper(II) oxide and hydrochloric acid.

15. Carbon dioxide and water are two of the three products from the reaction of sulfuric acid with sodium hydrogen carbonate.

G. Mixed Reactions

16. Hydrobromic acid reacts with potassium hydroxide.

17. Aluminum reacts with phosphoric acid.

18. Silver nitrate reacts with hydrosulfuric acid.

19. Phosphorus triiodide is formed from its elements.

20. Iron(II) chloride reacts with sodium phosphate.

21. Sugar, $C_{12}H_{22}O_{11}$, is burned in air.

22. Sugar, $C_{12}H_{22}O_{11}$, breaks down to carbon and water when heated.

23. Lithium hydroxide solution is the product of the reaction of lithium oxide and water.

24. Magnesium sulfate reacts with sodium hydroxide.

25. Chlorine reacts with a solution of sodium iodide.

26. Nickel hydroxide reacts with sulfuric acid.

Experiment 10

Work Page

27. Barium peroxide, BaO_2, decomposes into oxygen and barium oxide.

28. Ammonia is formed from its elements.

29. Butyl alcohol, C_4H_9OH, is oxidized completely.

30. Water is driven from copper sulfate pentahydrate, $CuSO_4 \cdot 5\,H_2O$, with heat.

31. Magnesium nitride is formed from its elements.

32. Sulfuric acid reacts with potassium nitrite.

Experiment 10

Report Sheet

Equation-Writing Exercise

Write the chemical equation for each reaction described below.

A. Combination Reactions

1. Diphosphorus trioxide is formed by direct combination of its elements.

2. Ammonia and sulfuric acid combine to form ammonium sulfate.

B. Decomposition Reactions

3. Ammonium nitrite decomposes into nitrogen and water.

4. When heated, potassium chlorate decomposes into oxygen and potassium chloride.

C. Complete Oxidation or Burning of Organic Compounds

5. Propane, C_3H_8, burns in air.

6. Acetaldehyde, CH_3CHO, is completely oxidized.

D. Oxidation–Reduction Reactions

7. Hydrogen is released when aluminum reacts with hydrochloric acid.

8. Magnesium reacts with silver nitrate solution.

E. Double Replacement Reactions

9. Barium carbonate precipitates from the reaction of barium chloride and sodium carbonate solutions.

10. Sulfuric acid reacts with calcium hydroxide.

11. Sodium iodate and silver nitrate solutions are combined.

12. Potassium fluoride reacts with hydrobromic acid.

13. Zinc hydroxide reacts with hydrochloric acid.

Experiment 10

Report Sheet

F. Other Reactions

14. Copper(II) chloride and water result from the reaction of copper(II) oxide and hydrochloric acid.

15. Carbon dioxide and water are two of the three products from the reaction of sulfuric acid with sodium hydrogen carbonate.

G. Mixed Reactions

16. Hydrobromic acid reacts with potassium hydroxide.

17. Aluminum reacts with phosphoric acid.

18. Silver nitrate reacts with hydrosulfuric acid.

19. Phosphorus triiodide is formed from its elements.

20. Iron(II) chloride reacts with sodium phosphate.

21. Sugar, $C_{12}H_{22}O_{11}$, is burned in air.

22. Sugar, $C_{12}H_{22}O_{11}$, breaks down to carbon and water when heated.

23. Lithium hydroxide solution is the product of the reaction of lithium oxide and water.

24. Magnesium sulfate reacts with sodium hydroxide.

25. Chlorine reacts with a solution of sodium iodide.

26. Nickel hydroxide reacts with sulfuric acid.

Experiment 10

Report Sheet

27. Barium peroxide, BaO_2, decomposes into oxygen and barium oxide.

28. Ammonia is formed from its elements.

29. Butyl alcohol, C_4H_9OH, is oxidized completely.

30. Water is driven from copper sulfate pentahydrate, $CuSO_4 \cdot 5\ H_2O$, with heat.

31. Magnesium nitride is formed from its elements.

32. Sulfuric acid reacts with potassium nitrite.

Mole Ratio for a Chemical Reaction

Performance Goals

11–1 Carry out a reaction between a measured amount of sodium hydrogen carbonate, $NaHCO_3$, and hydrochloric acid, HCl.

11–2 By weighing the solid product, derive the mole ratio in the reaction between the reactant and the product.

CHEMICAL OVERVIEW

In this experiment you will carry out a reaction where one of the products is a gas and the other is a solid.

$$NaHCO_3(s) + HCl(aq) \rightarrow NaCl(aq) + H_2O + CO_2(g) \qquad (11.1)$$

The above process will be carried out in a test tube. The "container" (see page 10) for this experiment, however, will be a beaker and the test tube. The beaker is necessary to hold the test tube on the balance in an upright position. Be sure to use the same beaker for each weighing.

To make sure that all the liquid has been evaporated, the solid residue should be "heated to constant mass." After the first heating, cooling, and weighing of the product, the test tube is heated, cooled, and weighed again. The two weighings should be within the plus-and-minus uncertainty of the balance used. If this is not the case, the heating procedure should be repeated again until the difference in mass is within the limits of the balance (about 0.005 for a milligram balance) between two consecutive weighings.

SAFETY PRECAUTIONS

Hydrochloric acid is corrosive. Avoid breathing the vapors and contact with skin. If acid should spill on you, *immediately* wash it off with *plenty* of water. *Extreme caution* should be exercised when heating the $NaHCO_3$ with HCl. Vigorous bubbling will occur, and when heating the contents of the test tube it will have a tendency to "shoot out." Be sure to move the test tube slowly in and out of the flame, while cautiously shaking the test tube. Always keep the test tube in a slanted position to expose the maximum area of the contents to heating. *Do not* point the open end of the test tube at anyone (including yourself!).

PROCEDURE

NOTE: Use a milligram balance for all measurements. Record data to the nearest 0.001 g.

A. Place a clean and dry 18 × 150-mm test tube in a beaker. Weigh this "container" and record its mass on the work page.

B. Add about 0.25–0.35 g of sodium hydrogen carbonate, $NaHCO_3$, to the test tube. Weigh the beaker, test tube, and contents. Record the mass.

C. One drop at a time, cautiously add 3 M hydrochloric acid, HCl, to the test tube. After each addition, gently shake the test tube until the reaction stops. Continue adding acid until there is no evidence of any further reaction (no more bubbles form). At this point no more solid should be present.

D. Evaporate all fluid by slowly moving the test tube in and out of the flame. Remember to keep the test tube slanted and shake it gently. Proceed until all liquid has evaporated.

E. Remove the test tube from the flame and test for water vapor by inverting a clean and dry test tube over the mouth of your test tube. If you see condensation, heat the test tube for an additional 5 minutes and test again. Alternatively, after the contents of the test tube appear dry, place the test tube in a 110–120°C oven and heat for 20 minutes.

F. Cool the test tube to room temperature. Place the test tube in the beaker used for earlier weighing and determine the mass.

CALCULATIONS

Using only numbers from your measurements, calculate by difference the initial mass of the sodium hydrogen carbonate. From the final weighing, calculate by difference the mass of the sodium chloride. From these quantities, using appropriate molar masses, calculate the number of moles of each compound.

$$\text{Moles} = \frac{\text{mass (g)}}{\text{molar mass (g/mole)}} \qquad (11.2)$$

Calculate the mole ratio of $NaHCO_3$ to NaCl. Compare this value to the theoretical value. Calculate the percent error (see Experiment 4).

Experiment 11

Advance Study Assignment

1. In the laboratory a student carried out the reaction

$$CaCO_3 + 2HCl \rightarrow CaCl_2 + H_2O + CO_2$$

The following data were collected:

Mass of container	48.365 g
Mass of container + $CaCO_3$	48.638 g
Mass of container + contents after evaporation of the liquid	48.664 g

Based on the data above, calculate:

a. Starting mass of $CaCO_3$

b. Mass of $CaCl_2$ obtained

c. Mole ratio of $CaCO_3$ to $CaCl_2$

2. List possible sources of error in this experiment.

Experiment 11

Work Page

Mass Data

Mass of container (g)	
Mass of container + NaHCO$_3$ (g)	
Mass of container + NaCl (g) 1st heating	
Mass of container + NaCl (g) 2nd heating	

Results

Mass of NaHCO$_3$ (g)	
Mass of NaCl (g)	

Calculations

1. Moles of NaHCO$_3$ and NaCl:

2. Mole ratio:

3. How does your experimental ratio compare to the theoretical? If they are different, explain what might have caused the difference.

4. Calculate the percent error.

Experiment 11

Report Sheet

Mass Data

Mass of container (g)	
Mass of container + NaHCO$_3$ (g)	
Mass of container + NaCl (g) 1st heating	
Mass of container + NaCl (g) 2nd heating	

Results

Mass of NaHCO$_3$ (g)	
Mass of NaCl (g)	

Calculations

1. Moles of NaHCO$_3$ and NaCl:

2. Mole ratio:

3. How does your experimental ratio compare to the theoretical? If they are different, explain what might have caused the difference.

4. Calculate the percent error.

Types of Chemical Reactions

Performance Goals

12–1 Carry out various chemical reactions.

12–2 Demonstrate that during chemical reactions mass is conserved.

CHEMICAL OVERVIEW

Chemical reactions can be classified as:

a. Combination or synthesis reactions in which two or more substances combine to form a single product.

b. Decomposition reactions, which are the opposite of the combination reactions, in that a compound breaks down into simpler substances.

c. Complete oxidation (burning) of organic compounds. In these reactions an organic compound reacts with oxygen yielding carbon dioxide, $CO_2(g)$, and water, $H_2O(g)$ or $H_2O(l)$.

d. Precipitation reactions, when the cation from one compound reacts with the anion of another compound yielding a solid product (precipitate). These reactions are also called double replacement or ion combination reactions since ions of the two reactants appear to change partners.

e. Oxidation–reduction reactions, during which one of the reactants gives off electrons (gets oxidized) and the other gains electrons (gets reduced).

f. Acid–base reactions, also called neutralization reactions, in which an acid reacts with a base yielding a salt and (usually) water.

In this experiment, we will carry out several different reactions, starting and ending with metallic copper. These reactions can be summarized as follows:

$$Cu \xrightarrow{HNO_3} Cu(NO_3)_2 \xrightarrow{NaOH} Cu(OH)_2 \xrightarrow{heat} CuO \xrightarrow{HCl} CuCl_2 \xrightarrow{Zn} Cu$$

SAFETY PRECAUTIONS

In this experiment you will use fairly concentrated acids and bases. When in contact with skin, most of these chemicals cause severe burns if not removed promptly. Always wear goggles when working with these chemicals. Reacting metal with nitric acid should *only* be carried out *in the hood*. Be careful when using a boiling water bath. Replenish the water from time to time as it becomes necessary.

PROCEDURE

1. Dissolution of Copper

A. Weigh a clean and dry 25 -mL Erlenmeyer flask on a milligram balance. Record this value on the work page.

B. Place about 100 mg of metallic copper (wire or granules) into the flask. Weigh the metal and flask to the milligram and record the mass.

C. *In the hood,* add 2 mL of 6 M nitric acid, HNO_3, to the flask and warm the contents on a hot plate. Brown vapors will form as the metal dissolves. Continue heating until no more brown fumes exist over the solution. Be sure not to evaporate all liquid. If needed, add two more milliliters of HNO_3. Allow the solution to cool to room temperature, then add 2 mL of deionized water.

2. Preparation of Copper(II) Hydroxide

To the solution prepared above, carefully add 6 M sodium hydroxide, NaOH, drop by drop, until the solution is basic to litmus (red paper turns blue). You can use magnetic stirring or swirl the contents of the flask while adding the NaOH. Do not dip the litmus paper directly into the solution. Instead, stir the solution with a glass rod and then touch the wet end of the rod to the paper. You should see a blue spot on the red paper when the solution is basic.

3. Preparation of Copper(II) Oxide

A. While stirring, heat the flask and its contents in a boiling water bath or on a magnetic stirring hot plate. In about 5 minutes, the blue $Cu(OH)_2$ will be converted to the black copper(II) oxide. If this does not occur, check your solution; it may not be basic enough. Swirl your solution and add more NaOH, then check with the litmus paper.

B. Allow the mixture to cool to room temperature. Remove the magnetic stirrer, if used. Rinse with a small amount of deionized water, collecting the rinse in the Erlenmeyer flask.

C. Set up a vacuum filtration apparatus using a small Büchner funnel (see Figure 31.2). Place a small filter paper into the funnel, moisten it with a small amount of deionized water, and start the vacuum. This will "seat" the filter paper and eliminates leakage around its edges. After this point, *do not* shut off the vacuum until the filtration is finished.

D. Transfer the black precipitate into the funnel, rinse the flask with 1 to 2 mL of deionized water, and pour into the funnel. The filtration may be a bit slow toward the end, due to small particles plugging up the filter paper pores. Wash the precipitate with 1 to 2 mL of deionized water. Discard the filtrate.

4. Converting Copper(II) Oxide to Copper(II) Chloride

Pour 6 mL of 6 M hydrochloric acid, HCl, into a 50-mL beaker. Using a spatula, transfer the black precipitate *and* the filter paper to the acid solution. Do not let the metal spatula come in contact with the acid. Stir the mixture with a glass stirring rod until the precipitate is completely dissolved. If needed, heat the solution on a hot plate. Remove the filter paper, using the glass rod, and rinse it with 1 to 2 mL of deionized water, adding the rinse to the green solution. *Do not* use metal forceps or tweezers because they will contaminate the solution. If some precipitate is stuck on the funnel, hold it over the beaker and rinse it with 1 to 2 mL of 6 M hydrochloric acid solution. Rinse the funnel with 1 to 2 mL of deionized water. This rinse should also be collected in the beaker.

5. Recovering the Metallic Copper

A. Weigh about 200 mg of zinc powder on a piece of preweighed weighing paper.

B. *In the hood,* **very** slowly, add a small amount of zinc powder to the copper(II) chloride solution. Stir after each addition. You will observe the formation of copper metal particles and vigorous evolution of hydrogen gas. This step is very critical, because too rapid formation of copper globules tends to enclose some of the unreacted zinc powder. This will result in an unrealistically high yield for the experiment.

C. Test for completeness of the reaction by adding 2 to 3 drops of your solution to 10 drops of concentrated ammonia, NH_3, in a small test tube. If a blue color appears, the reaction is not yet complete. Add a few more *small* portions of zinc powder and test again. Another indication that all the copper has been removed is the fact that the green solution turns colorless.

D. After the reaction is complete, add 5 mL of 3 M hydrochloric acid to the solution in the beaker and stir with a glass rod. This will hasten the removal of excess zinc present in your mixture. Metallic copper does not react with hydrochloric acid. Allow the solution to stand for 5 minutes, stirring occasionally.

E. Place a small funnel into a 250-mL Erlenmeyer flask or secure it with a clamp over a beaker. Weigh a piece of filter paper and place it in the funnel. First, pour the solution into the funnel, then transfer the solid copper. Use deionized water to rinse the beaker and be sure that all solid has been collected in the funnel. Wash the copper twice with 2-mL portions of deionized water.

F. Remove the filter paper and copper from the funnel, spread it out on a watch glass, and allow it to air dry. At the beginning of the next laboratory period weigh the copper and filter paper to the milligram and record the mass on the work page. Calculate the percentage of recovery (yield).

Experiment 12

Advance Study Assignment

1. Balance each of the following equations and classify the reactions:

   a. $Cu(s) + HNO_3(aq) \rightarrow Cu(NO_3)_2(aq) + H_2O + NO_2(g)$

   _____

   b. $Cu(NO_3)_2(aq) + NaOH(aq) \rightarrow Cu(OH)_2(s) + NaNO_3(aq)$

   _____

   c. $Cu(OH)_2(s) \rightarrow CuO(s) + H_2O$

   _____

   d. $CuO(s) + HCl(aq) \rightarrow CuCl_2(aq) + H_2O$

   _____

   e. $CuCl_2(aq) + Zn(s) \rightarrow Cu(s) + ZnCl_2(aq)$

   _____

2. Define a precipitation reaction.

3. If you started with 0.108 g of copper and at the end of the experiment you had recovered 0.099 g, calculate the percent recovery.

Experiment 12

Work Page

Mass Data

Mass of flask (g)	
Mass of flask + Cu (g)	
Mass of weighing paper (g)	
Mass of weighing paper + Zn (g)	
Mass of filter paper (g)	
Mass of filter paper + Cu (g)	

Results

Mass of Cu, initial (g)	
Mass of Zn (g)	
Mass of Cu, recovered (g)	
Percent recovery	

Show all your calculations below:

Classify each reaction as a double replacement, synthesis, decomposition, precipitation, neutralization, or oxidation–reduction reaction:

Part 1. _____

Part 2. _____

Part 3. _____

Part 4. _____

Part 5. _____

Experiment 12

Report Sheet

Mass Data

Mass of flask (g)	
Mass of flask + Cu (g)	
Mass of weighing paper (g)	
Mass of weighing paper + Zn (g)	
Mass of filter paper (g)	
Mass of filter paper + Cu (g)	

Results

Mass of Cu, initial (g)	
Mass of Zn (g)	
Mass of Cu, recovered (g)	
Percent recovery	

Show all your calculations below:

Classify each reaction as a double replacement, synthesis, decomposition, precipitation, neutralization, or oxidation–reduction reaction:

Part 1. _____

Part 2. _____

Part 3. _____

Part 4. _____

Part 5. _____

Qualitative Analysis of Some Common Ions

Performance Goals

13–1 Conduct tests to confirm the presence of known ions in a solution.

13–2 Analyze an unknown solution for certain ions.

CHEMICAL OVERVIEW

When we analyze an unknown solution, two questions come to mind: (1) what ions are present in the solution and (2) what is their concentration? The first question can be answered by performing a **qualitative analysis,** and the second by a **quantitative analysis**. These two broad categories are known collectively as **analytical chemistry**. In this experiment, you will perform a qualitative analysis.

The general approach to finding out what ions are in a solution is to test for the presence of each possible component by adding a reagent that will cause that component, if present, to react in a certain way. This method involves a series of tests, one for each component, carried out on separate samples of solution. Difficulty sometimes arises, particularly in complex mixtures, because one of the species may interfere with the analytical test for another. Although interferences are common, many ions in mixtures can usually be identified by simple tests.

In this experiment, you will analyze an unknown mixture that may contain one or more of the following ions in solution:

$$CO_3^{2-} \quad Cl^- \quad SCN^- \quad SO_4^{2-}$$
$$PO_4^{3-} \quad Cu^{2+} \quad Al^{3+}$$

First, you will perform the various tests designed to detect the presence of individual ions. Once you have observed these specific reactions, you will obtain the unknown solution from your instructor. Then, taking small portions of this solution, you will run each reaction again to determine which ions are present and which are absent.

This experiment is designed to test the behavior of only a few ions. More complex schemes are used for a more complete qualitative analysis.

SAFETY PRECAUTIONS AND DISPOSAL METHODS

In some tests you will be required to use fairly concentrated acids and bases. When in contact with skin, most of these chemicals cause severe burns if not removed promptly. Wear goggles when working with any of the reagents required in this experiment.

Discard all solutions in the container provided.

PROCEDURE

A boiling water bath is required for some of the tests you are to perform. Pour about 100 mL of deionized water into a 150-mL beaker and heat it to boiling. Maintain it at that temperature throughout the experiment, replenishing the water from time to time as it becomes necessary.

1. Test for the Carbonate Ion, CO_3^{2-}

Cautiously add about 10 drops of 1 M hydrochloric acid, HCl, to 10 drops of 1 M sodium carbonate, Na_2CO_3, in a small size test tube. Bubbles of colorless and odorless gas of carbon dioxide, CO_2, usually appear immediately, indicating the presence of the carbonate ion. If the bubbles are not readily apparent, warm the solution in the hot water bath and stir.

2. Test for the Sulfate Ion, SO_4^{2-}

Cautiously add about 10 drops of 1 M hydrochloric acid, HCl, to 10 drops of 0.5 M sodium sulfate, Na_2SO_4. Then add 3 to 4 drops of 1 M barium chloride, $BaCl_2$. A white, powdery precipitate of barium sulfate, $BaSO_4$, indicates the presence of the sulfate ions in the sample.

3. Test for the Phosphate Ion, PO_4^{3-}

Add 1 M nitric acid, HNO_3, to 10 drops of sodium phosphate, Na_3PO_4, until the solution is acidic. (Test by dipping a stirring rod into your solution and touching the wet rod to a strip of blue litmus paper. The solution is acidic when the blue litmus changes to a red color). Then add 5 drops of 0.5 M ammonium molybdate, $(NH_4)_2 MoO_4$, and heat the test tube in a hot water bath. A powdery, light yellow precipitate indicates the presence of the phosphate ion.

> **CAUTION**
>
> **The molybdate solution is yellow. Be sure you see a precipitate before you conclude that phosphate ions are present.**

4. Test for the Thiocyanate Ion, SCN^-

Add 10 drops of 3 M acetic acid, $HC_2H_3O_2$, to about 10 drops of 0.5 M potassium thiocyanate, KSCN, and stir with a glass stirring rod. Add 1 or 2 drops of 0.1 M iron(III) nitrate, $Fe(NO_3)_3$. The formation of a deep red color indicates the presence of thiocyanate ions.

5. Test for Chloride Ion, Cl^-

A. Add 5 drops of 1 M nitric acid, HNO_3, to about 10 drops of 0.5 M sodium chloride, NaCl. Add 2 or 3 drops of 0.1 M silver nitrate, $AgNO_3$. A white precipitate of silver chloride, AgCl, confirms the presence of silver ions.

B. If thiocyanate ion is present, it will interfere with this test, since it also forms a white precipitate with silver nitrate. If the sample contains

SCN⁻ ion, put 10 drops of the solution into a medium-size test tube and add 10 drops of 1 M nitric acid. Boil the solution gently until the volume is reduced to half. This procedure will oxidize the thiocyanate and remove the interference. Then perform the chloride ion test as previously explained.

6. Test for the Aluminum Ion, Al³⁺

Add 1 M ammonia, NH_3 dropwise to about 10 drops of 0.5 M aluminum chloride, $AlCl_3$, until the solution is basic (stir the solution with a glass stirring rod, then touch the wet rod to a strip of red litmus paper. When the solution is basic, the paper will turn blue). A white, gelatinous precipitate of aluminum hydroxide, $Al(OH)_3$, will form. Add 3 M acetic acid, $HC_2H_3O_2$, until the precipitate dissolves. Stir and add 2 drops of cathecol violet reagent. A blue solution indicates the presence of aluminum ions.

7. Test for the Copper(II) Ion, Cu²⁺

Add concentrated ammonia, NH_3, dropwise to about 10 drops of 0.5 M copper(II) sulfate, $CuSO_4$. The development of a deep blue color indicates the presence of copper(II) ions.

8. Analysis of an Unknown

When you have completed all of the tests, obtain an unknown from your instructor. *Be sure to record the unknown number.* Analyze your solution by using 10-drop portions of the unknown and then applying the various tests to separate portions. Substitute the unknown for the compound that contains the ion you are testing for. For example, if you are testing for the CO_3^{2-} ion, do not add Na_2CO_3; use your unknown in its place. The unknown will contain more than one ion, so your test for a given ion may be affected by the presence of others (see Step 5B).

Experiment 13

Advance Study Assignment

1. An unknown that might contain any of the seven ions studied in this experiment (but no other ions) has the following properties:

   a. On addition of 1 M HCl, bubbles form.

   b. When 0.1 M $BaCl_2$ is added to the acidified unknown, a white precipitate results.

   c. When 0.1 M $AgNO_3$ is added to the unknown, a clear solution results.

   On the basis of the preceding information, classify each of the following ions as present (P), absent (A), or undetermined (U) by the tests described:

   CO_3^{2-} _____ ; SO_4^{2-} _____ ; PO_4^{3-} _____ ; SCN^- _____ ;

   Cl^- _____ ; Al^{3+} _____ ; Cu^{2+} _____ ;

2. What is the difference between qualitative analysis and quantitative analysis?

Optional Assignment

Write net ionic equations for the reactions in this experiment in which the following ions are detected:

a. CO_3^{2-}: _____

b. SO_4^{2-}: _____

c. Cl^-: _____

Experiment 13

Work Page

Ions Tested	Observations (Known)	Unknown	
		Yes	No
CO_3^{2-}			
SO_4^{2-}			
PO_4^{3-}			
SCN^-			
Cl^-			
Al^{3+}			
Cu^{2+}			

Ions present _____

Experiment 13

Report Sheet

Ions Tested	Observations (Known)	Unknown	
		Yes	No
CO_3^{2-}			
SO_4^{2-}			
PO_4^{3-}			
SCN^-			
Cl^-			
Al^{3+}			
Cu^{2+}			

Ions present _____

Separation of Cations

Performance Goals

14–1 Prepare a solution containing five cations.

14–2 Separate and identify these cations using ion-specific reactions.

14–3 Analyze an unknown solution for certain cations.

CHEMICAL OVERVIEW

Qualitative analysis is a method that identifies the components of an unknown solution. In this experiment you will not only determine which ions are present, but you will also learn how to separate these ions in a mixture. This separation process is based on the specific chemical reactions each ion undergoes and also on the difference of solubility each compound has. You will use confirmatory tests also, to be sure that a particular ion is, in fact, present.

In this experiment you will learn to separate and identify five cations. The unknown solution may contain from one to five of the following ions:

$$Ag^+ \quad Pb^{2+} \quad Hg_2^{2+} \quad Fe^{3+} \quad Ni^{2+}$$

First, you will mix five solutions, each containing one of the above cations, to give a "known" solution. Then, following the given procedure, you will separate and identify each cation. When you are finished, you will be given an unknown and asked to analyze that solution and determine which cations are present.

SAFETY PRECAUTIONS AND DISPOSAL METHODS

In some tests you will be required to use fairly concentrated acids and bases that are very corrosive. If these come in contact with skin, be sure to rinse the exposed area with plenty of cold water. Wear goggles while performing this experiment. You will also be using a hot-water bath. Be careful not to spill the hot water. Handle test tubes in the hot-water bath by using a test-tube holder.

Discard all solutions in a waste container.

PROCEDURE

A. In a small beaker, mix 10 drops of each solution containing silver nitrate, $AgNO_3$, lead(II) nitrate, $Pb(NO_3)_2$, mercury(I) nitrate, $Hg_2(NO_3)_2$, iron(III) nitrate, $Fe(NO)_3$, and nickel(II) nitrate, $Ni(NO_3)_2$, to make a "known" solution. Stir well with a stirring rod.

B. Start heating some water for a water bath.

C. Pour about 1 mL of your known solution into a small test tube and add 10 drops of 6 M HCl. Mix thoroughly with a stirring rod (a small diameter rod works well in your small test tubes).

D. Put the test tube into a centrifuge, making sure that there is another test tube containing about the same amount of liquid in the opposite opening. Start the centrifuge and let it spin for about 20–30 seconds.

The precipitate should be on the bottom of the test tube and the liquid layer should be fairly clear. If this is not the case, centrifuge the mixture for 10 seconds longer. The precipitate contains $AgCl$, Hg_2Cl_2, and $PbCl_2$.

Pour the liquid into a small, clean test tube and save it for further tests.

NOTE: You will be using several test tubes, containing solutions from different steps and also precipitates from separation steps. It is extremely important that you label these test tubes, designating the step from which they were obtained and the ions they contain.

E. Add 1 to 2 mL of deionized water to the test tube containing the precipitate and stir it well with a glass rod. Centrifuge it for 20–30 seconds and discard the wash water.

F. Add about 2 mL of deionized water to the precipitate and, using a test-tube holder, place the test tube in a boiling water bath. (Be careful—the water level should be below the opening of the test tube!) Using a stirring rod, mix the contents of the test tube. Heat for about 3 minutes. The $PbCl_2$ will dissolve, but not the $AgCl$ or Hg_2Cl_2. Centrifuge the hot mixture and pour the hot liquid into another small test tube.

G. Allow the solution from Step F to cool. Then, add one drop of 6 M acetic acid, $HC_2H_3O_2$, and a few drops of 0.1 M potassium chromate, K_2CrO_4. If Pb^{2+} ions are present, $PbCrO_4$ will form, which is a bright yellow precipitate.

H. To the precipitate from Step F, add 1 mL of 6 M ammonia, NH_3, and stir thoroughly. Centrifuge the mixture and transfer the liquid into another test tube. If the remaining solid is gray or black, Hg_2^{2+} ions are present.

I. To the liquid from Step H, add 6 M nitric acid, HNO_3, slowly until a blue litmus paper turns red. Do not dip the paper into the solution; instead touch the wet stirring rod to the strip. If Ag^+ ions are present, the solution will turn cloudy, due to $AgCl$, which precipitates.

J. To the solution from Step D, add 6 M NH_3 until it is basic (i.e., a red litmus paper turns blue). Add 1 mL more NH_3 and stir. Centrifuge the mixture. The brown precipitate is $Fe(OH)_3$ and the nickel is in solution as a complex ion. Pour the solution into another test tube.

K. Dissolve the precipitate from Step J in about 0.5 mL of 6 M hydrochloric acid, HCl, and add 2 mL of deionized water. Stir, then add 2 drops of 0.5 M potassium thiocyanate, KSCN. If Fe^{3+} ions are present, a deep red solution will form.

L. To the solution from Step J, add a few drops of dimethylglyoxime reagent and stir. If Ni^{2+} ions are present, a rose-colored precipitate will form.

M. Obtain an unknown. Be sure to record the unknown number! The unknown may contain from one to five cations.

Perform Steps C through L exactly as you did for the known, except use the unknown where you used the known solution before. If, at some point, due to the absence of an ion, you do not get a precipitate or color change, proceed to the next step (i.e., if there is no precipitate, there is no need to centrifuge!).

Record your observations and list the ions present in your unknown.

Experiment 14

Advance Study Assignment

1. A solution contains Ag^+ ions and Pb^{2+} ions. Describe how you would separate them and what tests you would perform to verify that the specific ion is present.

2. An unknown that might contain any of the five cations studied in this experiment (but no other ions) has the following properties:

   a. On addition of HCl a white precipitate forms.

   b. The white precipitate is insoluble in hot water.

   c. After centrifuging the precipitate, the solution yields a brown precipitate when NH_3 is added.

   On the basis of the preceding information, classify each of the following ions as present (P), absent (A), or undetermined (U) by the tests described:

   Ag^+ _____ ; Pb^{2+} _____ ; Hg_2^{2+} _____ ; Fe^{3+} _____ ; Ni^{2+} _____

Experiment 14

Work Page

Ions Tested	Determining Reaction (Known)	Unknown	
		Yes	No
Ag^+			
Pb^{2+}			
Hg_2^{2+}			
Fe^{3+}			
Ni^{2+}			

Unknown number_____. Ions present _____

Experiment 14

Report Sheet

Ions Tested	Determining Reaction (Known)	Unknown	
		Yes	No
Ag^+			
Pb^{2+}			
Hg_2^{2+}			
Fe^{3+}			
Ni^{2+}			

Unknown number_____. Ions present_____

Molecular Models: A Study Assignment

Performance Goals

15–1 Write Lewis (electron dot) diagrams for molecules and ions formed by representative elements.

15–2 Predict the polarity of bonds and molecules formed by representative elements.

15–3 Predict bond angles and shapes of molecules and polyatomic ions.

15–4 Construct models for some covalently bonded species.

CHEMICAL OVERVIEW

Chemical bonds are the forces that hold atoms together in a compound. In this experiment we will study only covalently bonded species. A covalent bond is formed when a pair of electrons is shared by two atoms. Bonds in which the electrons are shared equally by the two nuclei are described as **nonpolar**. We expect to find nonpolar bonds whenever two identical atoms are joined, such as H_2 or Cl_2.

When different atoms are joined, the polarity of the bond depends on the electronegativity difference between the two elements. In the HF molecule, for example, the electron density is greater around the fluorine atom and the bond has a nonsymmetrical or nonuniform electron distribution. This type of a bond is referred to as **polar**.

We find experimentally that bonds formed between atoms that differ in electronegativity by 0.4 unit or less (such as the C—H bond) behave very much like pure nonpolar bonds and therefore may be classified as "essentially nonpolar." Bonds with an electronegativity difference greater than 1.7 are regarded as **ionic** bonds. (HF is an exception.) Electronegativity values for some common elements are listed in Table 15.1.

When determining whether a *molecule* is polar or not, its structure as well as the type of bonds in the molecule must be considered. If the molecule *has* at least one polar bond *and* is *nonsymmetrical*, it will be polar. On the other hand, even if the molecule has polar bonds, but is structurally symmetrical, the species will be nonpolar.* Obviously, if there are no polar bonds in a molecule, it will be nonpolar.

---

*The question of polarity does not apply to ionic species since they carry an overall charge.

Table 15.1 *Selected Electronegativity Values*

Element	Electroneg.	Element	Electroneg.
H	2.1	Si	1.8
B	2.0	P	2.1
C	2.5	S	2.5
N	3.0	Cl	3.0
O	3.5	Br	2.8
F	4.0	I	2.5

In this experiment you will be asked to determine the polarity of certain bonds and, after considering the geometry of the molecule, decide whether it is polar or nonpolar.

Lewis (Electron Dot) Diagrams

In most stable molecules or polyatomic ions, each atom tends to acquire a noble-gas structure by sharing electrons. This tendency is often referred to as the **octet rule**. One way to show the structure of an atom or a molecule is by using dots to represent the outermost s and p electrons (the so-called **valence electrons**). For the A group elements the number of valence electrons is the same as the group number in the periodic table.

Group	IA	IIA	IIIA	IVA	VA	VIA	VIIA	0
No. of valence electrons	1	2	3	4	5	6	7	8
Lewis structure	Li·	Ḃe	:Ḃ	:Ċ·	:N̈·	:Ö·	:F̈·	:N̈e:

In writing Lewis diagrams we usually do not attempt to show which atom the valence electrons come from; we simply indicate a shared pair of electrons by either two dots or a straight line connecting the atoms. Unshared pairs, also called lone pairs, are indicated by dots written around the elemental symbols. Example:

H_2O H—Ö—H NH_3 H—N̈—H with H below

Atoms in polyatomic ions are held together by covalent bonds. In the ions we will consider in this assignment, all atoms have a noble-gas structure.

Occasionally, too few electrons are available in a species to allow an octet to exist around each atom with only single bonds. In these instances multiple (double or triple) bonds will form.

Writing Lewis Structures

The Lewis diagrams of many species can be drawn by inspection. For more complex species, however, the following procedure is helpful:

1. **Draw a tentative diagram for the molecule or ion, joining atoms by single bonds. Place electron dots around each symbol except hydrogen so the total number of electrons for each atom is eight**. In some cases, only one arrangement of atoms is possible. In others, two or more structures may be drawn. Ultimately chemical or physical evidence must be used to decide which of the possible structures is correct. A few general rules will help you in drawing diagrams that are most likely to be correct:

   a. A hydrogen atom always forms one bond; a carbon atom normally forms four bonds.

   b. When several carbon atoms appear in the same molecule, they are often bonded to each other. In some compounds they are arranged in a closed loop; however, we will avoid such so-called cyclic compounds in this assignment.

   c. In compounds or ions having two or more oxygen atoms and one atom of another nonmetal, the oxygen atoms are usually arranged *around* the central nonmetal atom.

   d. In an oxyacid (hydrogen + oxygen + a nonmetal, such as H_2SO_4 or HNO_3), hydrogen is usually bonded to an oxygen atom, which is then bonded to the nonmetal: H—O—X, where X is a nonmetal.

2. **Count the electrons in your diagram**.

3. **Find the total number of valence electrons available**. For a molecule, this is the sum of the valence electrons contributed by each atom in the molecule. For a polyatomic ion, this total must be adjusted to account for the charge on the ion. An ion with a −1 charge will have one more electron than the number of valence electrons in the neutral atoms; a −2 ion, two more; a +1 ion, one less; and so forth.

4. **Compare the numbers in Steps 2 and 3. If they are the same, the diagram is complete. If they are different, modify the diagram with multiple bonds**. If your diagram has two electrons more than the number available, there will be one double bond. If the difference is four, there will be a triple bond or two double bonds. Multiple bonds should be used only when necessary, and then as few as possible should be used.

5. **The most common elements that can form a double bond are C, N, O, Si, P, and S**.

6. **The most common elements that can form a triple bond are C, N, Si, and P**.

7. **Halogens (i.e., Cl, Br, I) only form single bonds unless they are the center atom**.

There are some exceptions to the octet rule. It is not possible to write a Lewis diagram that has eight electrons around each atom if the total number of valence electrons is odd. Also, for compounds of Group IA, IIA, and IIIA elements, there are not enough electrons to satisfy the octet rule.

Electron Pair and Molecular Geometry

Several theories are used to explain the geometry (three-dimensional shape) of molecules. In this experiment we will use the valence shell electron pair repulsion (VSEPR) theory. According to this theory, electrostatic repulsion arranges the electron pairs surrounding an atom so that they are as far from each other as possible. This arrangement is the **electron pair geometry**. The molecular geometry, the arrangement of *atoms* around the central atom, is a direct result of the electron pair geometry.

Let us consider cases in which the central atom is surrounded by 2, 3, or 4 electron pairs.

1. **Two pairs of electrons**. Two pairs of electrons around a central atom are farthest from each other when they are on opposite sides of that atom. The electron pairs and the central atom are on the same straight line. The electron pair geometry is **linear**. Both electron pairs bond atoms to the central atom, so all three atoms are also on the same line. The molecular geometry is linear too.

2. **Three pairs of electrons**. Three electron pairs will be farthest apart when they are directed toward the corners of an equilateral triangle with the central atom at its center. The atom and all electron pairs are in the same plane, so the electron pair geometry is called **trigonal planar,** or **planar triangular**. If all three electron pairs are bonding pairs, the molecule is also planar triangular, with 120° bond angles. If one of the electron pairs is a lone pair, the molecular geometry is **bent** or **angular**.

3. **Four pairs of electrons**. Four pairs of electrons in three dimensions are farthest apart when located at the corners of a tetrahedron with the central atom in its center. If all four electron pairs are bonding pairs, as in CH_4, the central carbon atom is in the center of the tetrahedron and the four hydrogen atoms are at its corners. Both the electron pair and molecular geometries are **tetrahedral**. The bond angles are 109°.

   If there are three bonding pairs and one lone pair, as in the NH_3 molecule, VSEPR still predicts a tetrahedral electron pair geometry with a tetrahedral angle for the bonding electron pairs. (In fact, the angle is slightly less than tetrahedral, but we will call it a "tetrahedral" angle.) The molecule has the shape of a low pyramid; its geometry is **trigonal pyramidal**. Two bonding pairs and two lone pairs around the central atom again yield close to a tetrahedral angle. The three-atom molecule has a **bent** geometry.

These electron pair and molecular geometries are summarized in Table 15.2.

Multiple Bonds

Bond angles in species that contain multiple bonds indicate that the electrons in the multiple bond behave as a single pair of electrons according to VSEPR. Thus a structure such as

has a planar triangular geometry around the C atom with 120° bond angles.

Table 15.2 *Electron Pair and Molecular Geometry*

Total No. Electron Pairs	Shared	Unshared	Bond Angle	Electron Pair Geometry	Molecular Geometry	Example
2	2	0	180°	Linear	Linear	BeF_2
3	3	0	120°	Trig. planar	Trig. planar	BF_3
3	2	1	120°	Trig. planar	Angular (bent)	NO_2^-
4	4	0	109°	Tetrahedral	Tetrahedral	CCl_4
4	3	1	109°	Tetrahedral	Trigonal pyramidal	NH_3
4	2	2	109°	Tetrahedral	Bent	H_2O

PROCEDURE

Obtain a molecular model kit. Draw a Lewis diagram for HCl, the first item in the table on the work page. From your diagram, and using Table 15.2 as a guide, fill in all the blanks for HCl. Then build a model of an HCl molecule. Use the model to verify the geometry you predicted. If the model and your prediction do not agree, find out why. Do not proceed to H_2O until you thoroughly understand the HCl structure.

Follow the same procedure with H_2O, the next item in the table. This time you must predict a bond angle too. Your model should confirm your predictions. Again, do not proceed to NH_3 until you thoroughly understand H_2O.

Proceed in a similar manner for all species shown in the table. From a learning standpoint, it is important that you complete each species, including the model, before you proceed to the next.

Experiment 15

Advance Study Assignment

1. Define electron pair and molecular geometry. When are these the same and when are they different? Give an example.

2. Draw Lewis structures for CO_2, SO_2, and NO_3^-.

3. Give the electron pair geometry and the molecular geometry of the three species from Question 2, according to VSEPR.

4. Are CO_2 and SO_2 polar or nonpolar molecules? Explain your reasoning.

Experiment 15

Work Page

Instructions: For each species listed below, draw the Lewis structure first and then complete the rest of the information requested for that species. "Build" the molecule or ion with your model kit. Verify from your model the geometry you predicted. Fill out each line and build the model before proceeding to the next line. If there are more than two atoms as "center," you can only deduce the geometry separately for each one. Continue this procedure for each species listed here.

Formula	Lewis Structure	Electron Pair Geometry	Molecular Geometry	Bond Angle	Polar	Nonpolar
HCl		X		X		
H_2O						
NH_3						
BCl_3						
CO_2						
IO_2^-					X	X
PO_4^{3-}					X	X

Formula	Lewis Structure	Electron Pair Geometry	Molecular Geometry	Bond Angle	Polar	Nonpolar
C_2H_6						
C_2H_4						
C_2H_2		Around O	Around O			
C_2H_5OH						
HCHO		Around O	Around O			
CH_3—O—CH_3		Around C	Around C			
HCOOH		Around single-bonded O				
CH_3COOH						

Experiment 15

Report Sheet

Instructions: For each species listed below, draw the Lewis structure first and then complete the rest of the information requested for that species. "Build" the molecule or ion with your model kit. Verify from your model the geometry you predicted. Fill out each line and build the model before proceeding to the next line. If there are more than two atoms as "center," you can only deduce the geometry separately for each one. Continue this procedure for each species listed here.

Formula	Lewis Structure	Electron Pair Geometry	Molecular Geometry	Bond Angle	Polar	Nonpolar
HCl		X		X		
H_2O						
NH_3						
BCl_3						
CO_2						
IO_2^-					X	X
PO_4^{3-}					X	X

Formula	Lewis Structure	Electron Pair Geometry	Molecular Geometry	Bond Angle	Polar	Nonpolar
C_2H_6						
C_2H_4						
C_2H_2						
C_2H_5OH		Around O	Around O			
HCHO						
CH_3—O—CH_3		Around O	Around O			
HCOOH		Around C	Around C			
CH_3COOH		Around single-bonded O	Around single-bonded O			

Molar Volume of a Gas

Performance Goals

16–1 Determine the molar volume of a gas at room temperature and pressure from experimental data.

CHEMICAL OVERVIEW

The molar volume of a gas is the number of liters occupied by 1 mole of the gas. It is the purpose of this experiment to measure the volume occupied by a known number of moles of hydrogen. Dividing the volume in liters by the number of moles gives the molar volume in liters per mole.

The volume occupied by a fixed quantity of gas depends on temperature and pressure. Because this entire experiment will be carried out at room temperature—the temperature at which the molar volume is to be found—this variable may be disregarded in your calculations.

The quantitative relationship between pressure and volume of a fixed amount of gas at constant temperature, known as Boyle's Law, can be expressed mathematically as

$$PV = \text{constant} \qquad \text{or} \qquad P_1V_1 = P_2V_2 \qquad (16.1)$$

where P is pressure, V is volume, and subscripts 1 and 2 refer to the first and second measurements of the gas sample.

In a mixture of gases, each component exerts a certain amount of pressure of its own. This pressure, called the **partial pressure**, is defined as that pressure which the component would exert if it alone occupied the total volume at the given temperature. Mathematically,

$$P = p_1 + p_2 + p_3 + \dots \qquad (16.2)$$

where P is the total pressure of the mixture and $p_1, p_2, p_3, \dots$ are the partial pressures of compounds 1, 2, 3, etc. The relationship is known as Dalton's Law of Partial Pressures.

When a confined gas is generated in contact with water, the gas becomes saturated with water vapor, producing a "wet" gas—a gaseous mixture. The partial pressure of water vapor is a function of temperature and can be found in tables. Table 16.1 contains a partial list of water vapor pressures at various temperatures.

In this experiment, hydrogen will be generated by the reaction of a measured mass of metallic magnesium with excess hydrochloric acid. The volume of gas produced will be measured in a gas measuring tube, called a *eudiometer*. Using Boyle's and Dalton's laws, the measured volume will be

Table 16.1 *Water Vapor Pressure in Torr*

Temperature (°C)	Vapor Pressure (torr)	Temperature (°C)	Vapor Pressure (torr)
17	14.5	23	21.1
18	15.5	24	22.4
19	16.5	25	23.8
20	17.5	26	25.2
21	18.6	27	26.7
22	19.8	28	28.3

corrected to obtain the volume of *dry* hydrogen at room temperature. The number of moles of hydrogen will be calculated by stoichiometry from the mass of magnesium that reacted. Dividing volume by moles gives the molar volume at room temperature.

A word about significant figures: Ideally you should weigh your magnesium on an analytical balance to the nearest 0.0001 g. Some schools do not have such balances available to students and you can use a milligram balance that measures to the nearest 0.001 g. Ordinarily, strict application of the rules of significant figures makes this the measurement that determines the number of significant figures in your final result. It is a fact, however, that more sophisticated—and more correct—ways of handling measurement uncertainty in this experiment yield an answer with one extra digit. Therefore, if you use a milligram balance, report your final result by going one digit beyond what normal significant figure rules allow.

SAMPLE CALCULATIONS

Example 1

Calculate the pressure of dry N_2 at 26°C if it is collected over water and the total pressure is 747 torr. Water vapor pressure at this temperature is 25.2 torr.

$$P = p_{N_2} + P_{H_2O} \quad ; \quad p_{N_2} = P - p_{H_2O}$$

Substituting,

$$p_{N_2} = 747 \, \text{torr} - 25 \, \text{torr} = 722 \, \text{torr}$$

Example 2

What volume would the dry N_2 of Example 1 occupy at 762 torr pressure if the experimentally measured volume was 80.5 mL?

Substituting into Equation 16.1,

$$P_1V_1 = P_2V_2 \quad ; \quad V_2 = \frac{P_1V_1}{P_2}$$

$$V_2 = 722 \, torr \times \frac{80.5 \, mL}{762 \, torr} = 76.3 \, mL$$

Example 3

If 0.0769 g of magnesium produced the gas in Examples 1 and 2, how many moles of hydrogen were produced? $Mg + 2 \, HCl \rightarrow H_2 + MgCl_2$.

$$0.0769 \, g \, Mg \times \frac{1 \, mole \, Mg}{24.3 \, g \, Mg} \times \frac{1 \, mole \, H_2}{1 \, mole \, Mg} = 0.00317 \, mole \, H_2$$

Example 4

Using the results of Examples 1 to 3, find the molar volume of hydrogen at room conditions.

$$\frac{0.0763 \, L \, H_2}{0.00317 \, mole \, H_2} = 24.1 \, L \, H_2/mole$$

SAFETY PRECAUTIONS AND DISPOSAL METHODS

Be careful when handling concentrated hydrochloric acid. If it comes into contact with your skin, wash it with plenty of water. Wear goggles when performing this experiment.

The hydrochloric acid solution obtained in the eudiometer can be poured down the drain or disposed of as specified by your instructor.

PROCEDURE

NOTE: Record the mass of magnesium in grams to the nearest 0.0001 g if you use an analytical balance, or 0.001 g if you use a milligram balance (see Chemical Overview). Record volume measurements in milliliters to the nearest 0.1 mL. Record atmospheric pressure in torr to the nearest torr and temperature to the nearest °C.

A. Obtain a strip of magnesium that weighs between 0.072 and 0.082 g. Clean both sides by rubbing with steel wool. Weigh the magnesium, preferably on an analytical balance, and record the mass to the nearest 0.0001 g. (If you use a milligram balance, assume that calculations from this measurement are good to one digit beyond the number normally allowed by the rules of significant figures. See the Chemical Overview.)

B. Wad the magnesium into a small irregularly shaped ball. Do not fold it neatly or pack it tightly, because the reaction requires that the acid be able to reach all surfaces of the magnesium. Wrap this ball in a small "cage" made of copper wire. Be sure that no large "holes" are present

Figure 16.1

Apparatus for the volume measurement of a gas.

in your cage, or pieces of magnesium will escape during the reaction. Leave a 10- to 12-cm piece of wire extending from the cage (see Figure 16.1).

C. Fill a 500-mL cylinder to the top with tap water and stand it in a large beaker, tray, or the sink to catch the overflow. In selecting a place, read ahead to Step F. Make sure you can perform that step without knocking the cylinder over.

D. Carefully pour 10 mL of 6 M hydrochloric acid, HCl, into a 100-mL eudiometer, or gas measuring tube. Then, holding the tube at an angle, slowly pour deionized water down the side of the tube. When the water nears the top of the eudiometer, hold it vertically and continue to add water until the tube is totally full.

E. Carefully lower the copper-wire-wrapped magnesium ball into the eudiometer until it is about 5 or 6 cm into the tube, but not so deep that it extends into the calibrated portion of the eudiometer. Place a one-hole rubber stopper into the eudiometer, securing the straight piece of copper wire to the wall of the tube. Water will come out the hole in the stopper when it is placed into the eudiometer, so be ready for it.

F. Place your finger over the hole of the stopper, invert the eudiometer, and quickly lower it into the water-filled cylinder. When the stopper is beneath the surface of the water, remove your finger. Move the eudiometer and cylinder to a clamp mounted on a ring stand, as shown in Figure 16.1; a buret clamp is ideal for this purpose.

G. Allow the reaction to proceed until no more bubbles are formed.

CAUTION

There is apt to be more overflow of water from the cylinder, so be prepared. Tap the sides of the eudiometer gently to dislodge any trapped bubbles from the wire cage.

H. Carefully adjust the eudiometer vertically until the water level inside the tube is at the same level as the water in the cylinder. At this point the total pressure inside the tube is the same as atmospheric pressure.

I. Read and record the volume of the gas generated.

J. Read and record the atmospheric pressure and temperature in the laboratory.

K. Discard the wire cage into a wastebasket or recycle it, as instructed. Carefully dispose of the liquid into the sink or into a waste container. (The liquid contains hydrochloric acid, remember?)

Experiment 16

Advance Study Assignment

1. Calculate the number of moles of gas generated when 0.165 g of aluminum reacts with hydrochloric acid.

$$2Al(s) + 6HCl(aq) \rightarrow 2AlCl_3(aq) + 3H_2(g)$$

2. Calculate the pressure of dry O_2 if the total pressure of O_2 generated over water is measured to be 638 torr and the temperature is 22.0°C. $p_{H_2O} = 19.8$ torr.

3. If the volume of the O_2 sample in Question 2 was 56.2 mL, what volume would the dry O_2 occupy at 755 torr (assume temperature unchanged)?

Experiment 16

Work Page

Data

Run	1	2	3
Mass of Mg strip (g)			
Room pressure (torr)			
Room temperature (°C)			
Volume of wet H_2 gas (mL)			
Water vapor pressure (torr)			

Calculations

Moles of Mg			
Moles of H_2 gas			
Partial pressure of dry H_2 (torr)			
Volume of dry H_2 at room pressure (mL)			
Molar volume of dry H_2 at room temperature and pressure (L/mole)			

Show setups for all calculations below:

Optional

From *your* data obtained in this experiment, calculate the value of the universal gas constant R. Give proper unit.

Calculate the percent error (look up the known value).

Experiment 16

Report Sheet

Data

Run	1	2	3
Mass of Mg strip (g)			
Room pressure (torr)			
Room temperature (°C)			
Volume of wet H_2 gas (mL)			
Water vapor pressure (torr)			

Calculations

Moles of Mg			
Moles of H_2 gas			
Partial pressure of dry H_2 (torr)			
Volume of dry H_2 at room pressure (mL)			
Molar volume of dry H_2 at room temperature and pressure (L/mole)			

Show setups for all calculations below:

Optional

From *your* data obtained in this experiment, calculate the value of the universal gas constant R. Give proper unit.

Calculate the percent error (look up the known value).

Molar Mass Determination by Freezing-Point Depression

Performance Goals

17–1 Measure the freezing point of a pure substance.

17–2 Determine the freezing point of a solution by graphical methods.

17–3 Knowing the molal freezing-point constant and having determined experimentally the mass of an unknown solute, the mass of a known solvent, and the freezing-point depression, find the molar mass of the solute.

CHEMICAL OVERVIEW

When a solution is prepared by dissolving a certain amount of solute in a pure solvent, the properties of the solvent are modified by the presence of the solute. Changes in such properties as the melting point, boiling point, and vapor pressure are found to be dependent on the number of solute molecules or ions in a given amount of solvent. The nature of the solute particles (molecules or ions) is not important; the governing factor is the *relative number of particles*. Properties that are dependent only on the concentration of solute particles are referred to as **colligative properties**.

If a pure liquid is cooled, the temperature will decrease until the freezing point is reached. With continued cooling, the liquid gradually freezes. As long as liquid and solid are *both* present, the temperature will remain constant. When all the liquid is converted to solid, the temperature will drop again.

A typical cooling curve for a pure solvent is shown in Figure 17.1. The dip below the freezing point is the result of supercooling, an unstable situation in which the temperature drops below the normal freezing point until crystallization begins. Supercooling may or may not occur in any single freezing process and will probably vary in successive freezings of the same sample of pure substance. As soon as crystal formation begins, the temperature rises to the normal freezing point and remains constant until all of the liquid is frozen.

The freezing point of a solution is always lower than that of the pure solvent. The difference between the freezing points of the solvent and solution is the **freezing-point depression**. This difference is proportional to the molal concentration of solute (see Equation 17.1). During the freezing of a solution, it is the solvent that freezes. Therefore, the solvent is gradually removed from the solution as the freezing progresses, leaving behind an increasingly concentrated solution. Because of this concentration increase, the freezing point drops, producing a solution freezing curve such as that in Figure 17.2. Again, the supercooling effect may or may not be observed in any single freezing experiment.

Colligative properties may be used to find the molar mass of an unknown substance. It is known, for example, that 1 mole of any molecular substance (one that does not produce ions in solution), when dissolved in 1 kg of water, lowers its freezing point by 1.86°C and raises its boiling point by 0.52°C. The freezing-point depression or boiling-point elevation for a 1.0 molal solution of a molecular substance will be the same regardless of the nature of the substance, as long as no ionization takes place. In general, the relationship between freezing-point depression and concentration of solute can be expressed as

$$\Delta T_F = K_F m \tag{17.1}$$

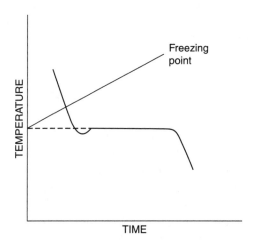

Figure 17.1
Cooling curve of a pure liquid

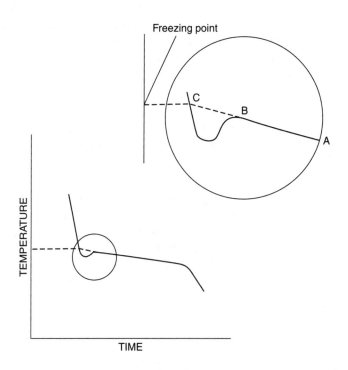

Figure 17.2
Cooling curve of a solution

where ΔT_F is the freezing-point depression in °C, K_F is the molal freezing-point constant of the pure solvent, and m is the molality of the solution (moles of solute per kg of solvent).

The solvent in this experiment is naphthalene ($K_F = 6.9°C/m$). You will determine its freezing point experimentally with *your* thermometer. (The literature value of the freezing point is 80.2°C.) You will then prepare a solution from a weighed quantity of naphthalene and a weighed quantity of unknown solute, and determine experimentally its freezing point. When you substitute ΔT_F into Equation 17.1, the molality of the solution can be calculated.

Molality, by definition, is moles of solute per kilogram of solvent. This may be expressed mathematically as

$$m = \frac{\text{mole solute}}{\text{kg solvent}} = \frac{\text{g solute/MM}}{\text{kg solvent}} \qquad (17.2)$$

Knowing the mass of the solute and the mass of the solvent used, the molar mass (MM) can be calculated if the molality (m) is determined experimentally.

SAMPLE CALCULATIONS

Example 1

Calculate the molal concentration of a solution of 3.45 g of compound A (MM =120 g/mole) dissolved in 50.0 g of solvent.

Converting the mass of solvent to kg and substituting the given values into Equation 17.2,

$$m = \frac{3.45 \text{ g solute/120 g solute/mole}}{0.050 \text{ kg solvent}}$$
$$= \frac{3.45 \text{ g solute}}{0.050 \text{ kg solvent}} \times \frac{1 \text{ mole solute}}{120 \text{ g solute}} = 0.575 \text{ mole/kg solvent}$$

Example 2

Calculate the molar mass of an unknown substance if a solution containing 2.37 g of unknown in 20.0 g of water freezes at −1.52°C. K_F for water = 1.86°C/m.

The normal freezing point of water is 0.00°C, and that of the solution is −1.52°C. The freezing-point depression therefore is

$$\Delta T_F = 0.00°C - (-1.52°C) = 1.52°C$$

Solving Equation 17.1 for m and substituting,

$$m = \frac{\Delta T_F}{K_F} = \frac{1.52°C}{1.86°C/m} = 0.817 \frac{\text{mole solute}}{\text{kg solvent}}$$

Now you can calculate the molar mass using Equation 17.2:

$$0.817 \frac{\text{mole solute}}{\text{kg solvent}} = \frac{2.37 \text{ g solute/MM}}{0.0200 \text{ kg solvent}}$$

After rearranging to obtain MM,

$$MM = \frac{2.37 \text{ g solute}}{0.0200 \text{ kg solvent} \times 0.817 \dfrac{\text{mole solute}}{\text{kg solvent}}} = 145 \text{ g/mole}$$

SAFETY PRECAUTIONS AND DISPOSAL METHODS

In this experiment you will be handling various organic chemicals. Always use a spatula; do not handle crystals by hand. When using organic liquids, do not breathe their vapors.

Dispose of excess reagents as directed by your instructor. Do not use organic liquids near an open flame.

PROCEDURE

NOTE: Record all mass measurements in grams to the nearest 0.01 g. Record all temperature measurements in degrees Celsius to the nearest 0.1 °C.

1. The Freezing Point of Naphthalene

A. Assemble the apparatus shown in Figure 17.3 using a thermometer graduated in tenths of degrees Celsius. A circular stirrer may be formed from No. 18 wire. A 600-mL beaker half full of water may be used as the water bath.

Figure 17.3

Apparatus for determining freezing point

B. Weigh an empty test tube to ±0.01 g on a centigram balance. Add about 10 g of naphthalene and weigh again.

C. Insert the thermometer in the test tube and place the unit in the water bath. Heat the water bath until the naphthalene completely melts. Be sure the entire thermometer bulb is submerged in the molten naphthalene and is not touching the bottom or walls of the test tube.

D. Discontinue heating, remove the water bath, and allow the liquid to cool slowly. Beginning at about 90°C, record the temperature to ±0.1 °C every 30 seconds for 10 minutes. Stir continuously during this period by raising and lowering the wire stirrer.

2. The Freezing Point of the Solution

A. Place a 1.0- to 1.3 g sample of unknown solute into a test tube and weigh them to the nearest 0.01 g on a centigram balance. Transfer the unknown to the test tube containing naphthalene (Part 1). Again weigh the empty test tube with any powder of the unknown that may not have been transferred, to determine the mass of the solute dissolved.

B. Heat the mixture in the water bath until all the naphthalene is melted and the unknown is completely dissolved. With constant stirring, determine the freezing point of the solution exactly the same way as you determined the freezing point of the pure naphthalene.

C. If time permits, remelt the solution and repeat the freezing cycle.

3. Clean-Up

To clean your equipment, melt the solution in the test tube and pour it onto a paper towel folded to several thicknesses. The test tube may be cleaned with hexane, acetone, or some other suitable solvent.

> **CAUTION**
>
> Do *not* pour melted naphthalene solutions into the sink. They freeze and clog drains. Many cleaning solvents are flammable and should not be used near an open flame. Clean the equipment in a hood. Do not breathe vapors.

FREEZING-POINT DETERMINATION

Estimate to ±0.1 °C the freezing point of pure naphthalene from your data. According to Figure 17.1, the freezing point of a pure substance is the constant temperature at which the substance freezes. With due allowance for supercooling, if any, this temperature should be apparent from the table. Record it on the work page.

To estimate the freezing temperature of the solution of the unknown in naphthalene, plot the temperature vs. time on the graph paper provided. The freezing point of the solution *of the concentration you prepared* is the point where it first begins to freeze if there is no supercooling. If supercooling does occur, the freezing point may be estimated from the graph by extrapolating the curve during freezing (from A to B in Figure 17.2) back to its intersection with the cooling line for the liquid (point C). The temperature corresponding to point C is the temperature at which freezing would have begun in the absence of supercooling—the freezing point of your solution at its initial concentration. Record the freezing temperature of the solution on the work page.

CALCULATIONS

From the observed freezing points of pure naphthalene and the solution, calculate the freezing-point depression. Record ΔT_F in the Results table on the work page.

Using Equation 17.1 and the molal freezing-point constant of naphthalene ($K_F = 6.9°C/m$), find the molality of your solution. Record m in the Results table on the work page.

Determine and record the kilograms of naphthalene and grams of solute in your solution.

Using Equation 17.2 and the results already determined, calculate and record the molar mass of your unknown.

Experiment 17

Advance Study Assignment

1. Define colligative properties.

2. Calculate the molality of the solution obtained when 10.3 g of sugar, $C_6H_{12}O_6$, are dissolved in 275 g of water.

3. Calculate the freezing-point depression for a solution prepared by dissolving 25.0 g of C_8H_{18} in 175 g of benzene. ($K_F = 4.90°C/m$ for benzene.)

4. If 2.6 g of unknown solute are dissolved in 28 g of water, the freezing-point depression is 2.7°C.

   a. Calculate the molality of the solution if $K_F = 1.86°C/m$ for water.

   b. Calculate the molar mass of the solute.

Experiment 17

Work Page

Data

Time-Temperature Readings (Read to ± 0.1 °C)

Time (Minutes)	Naphthalene	Naphthalene + Unknown			
0					
$\frac{1}{2}$					
1					
$1\frac{1}{2}$					
2					
$2\frac{1}{2}$					
3					
$3\frac{1}{2}$					
4					
$4\frac{1}{2}$					
5					
$5\frac{1}{2}$					
6					
$6\frac{1}{2}$					
7					
$7\frac{1}{2}$					
8					
$8\frac{1}{2}$					
9					
$9\frac{1}{2}$					
10					

Experiment 17

Work Page

Data

Mass of test tube (g)			
Mass of test tube and naphthalene (g)			
Mass of test tube and unknown _____ (g)			
Mass of test tube after transfer (g)			
Freezing point of naphthalene (°C)			
Freezing point of solution (from curve) (°C)			

Results

Freezing-point depression, ΔT_F (°C)			
Molality, m (mole solute/kg solvent)			
Mass of naphthalene (g)			
Mass of unknown _____ (g)			
Molar mass of unknown (g/mole)			

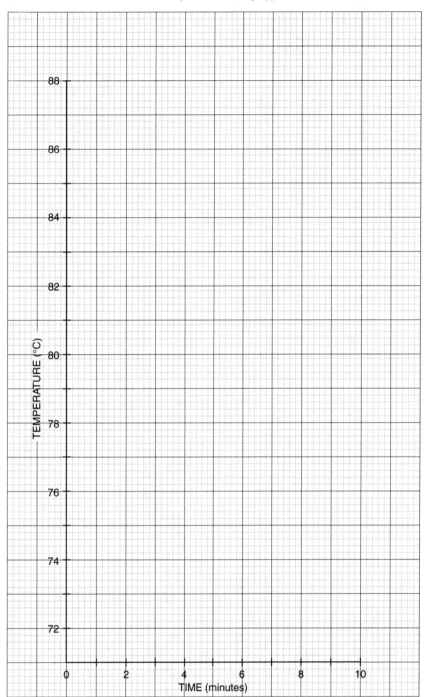

Cooling curve of solution

Experiment 17

Report Sheet

Data

Time-Temperature Readings (Read to $\pm 0.1\ ^\circ$C)

Time (Minutes)	Naphthalene	Naphthalene + Unknown			
0					
$\frac{1}{2}$					
1					
$1\frac{1}{2}$					
2					
$2\frac{1}{2}$					
3					
$3\frac{1}{2}$					
4					
$4\frac{1}{2}$					
5					
$5\frac{1}{2}$					
6					
$6\frac{1}{2}$					
7					
$7\frac{1}{2}$					
8					
$8\frac{1}{2}$					
9					
$9\frac{1}{2}$					
10					

Experiment 17

Report Sheet

Data

Mass of test tube (g)			
Mass of test tube and naphthalene (g)			
Mass of test tube and unknown _____ (g)			
Mass of test tube after transfer (g)			
Freezing point of naphthalene (°C)			
Freezing point of solution (from curve) (°C)			

Results

Freezing-point depression, ΔT_F (°C)			
Molality, m (mole solute/kg solvent)			
Mass of naphthalene (g)			
Mass of unknown _____ (g)			
Molar mass of unknown (g/mole)			

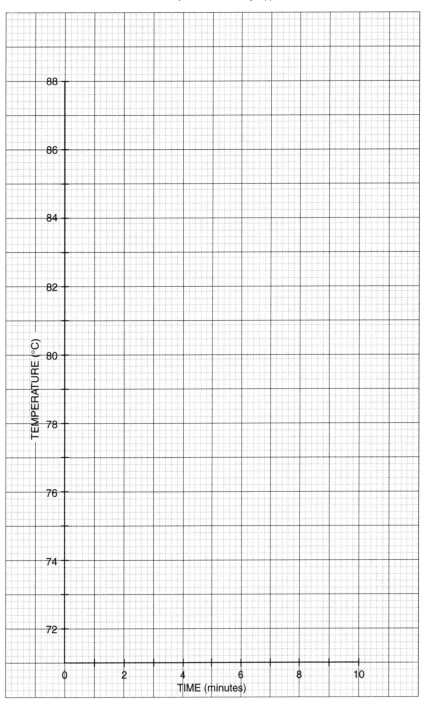

Cooling curve of solution

The Conductivity of Solutions: A Demonstration

Performance Goals

18–1 Describe how the conductivity of a solution may be tested.

18–2 Basing your decision on conductivity observations, classify substances as strong electrolytes, weak electrolytes, or nonelectrolytes.

18–3 Explain the presence or absence of conductivity in an aqueous solution.

CHEMICAL OVERVIEW

Solutions of certain substances are conductors of electricity. The conductance is due to the presence of charged species (ions) that are free to move through the solution. If two metal strips, called *electrodes*, are connected to a current source (such as a battery or regular wall plug) and then immersed into a conducting solution, current will flow through the system. Ions will be attracted to the oppositely charged electrodes; i.e., positively charged ions, called *cations*, will flow to the negatively charged electrode (the *cathode*), and negatively charged *anions* will be attracted to the positively charged *anode*. Ions are already present in solid *ionic compounds*, but they are in fixed positions and hence cannot move. They are simply "released" from the crystal when the compound dissolves, and thereby become mobile. An example is sodium chloride, NaCl, which may be written $Na^+Cl^-(s)$ to emphasize its character as an ionic solid:

$$Na^+Cl^-(s) \xrightarrow{\ H_2O\ } Na^+(aq) + Cl^-(aq) \qquad (18.1)$$

With some *molecular compounds* (containing neutral molecules, instead of ions, in their pure state), ions are formed by reaction of the solute with water. For example, hydrogen chloride gas, $HCl(g)$, reacts with water and yields hydrochloric acid:

$$HCl(g) + H_2O(l) \rightarrow H_3O^+(aq) + Cl^-(aq) \qquad (18.2)$$

In dilute hydrochloric acid, this reaction is virtually complete, and no appreciable number of neutral HCl molecules are present.

Solutions that contain a large number of ions are *good conductors*. The solutions, and the solutes that produce them, are called **strong electrolytes.** Hydrochloric acid is a strong acid. As stated above, its ionization is essentially complete, yielding a large number of ions. Similarly, if you dissolve solid potassium hydroxide, KOH, in water, ionization will be complete and the resultant solution will be a good conductor. There are relatively few strong acids and bases. All *soluble* salts completely ionize when dissolved in water and hence yield strongly conducting solutions.

When acetic acid is dissolved in water, an equilibrium is reached:

$$CH_3COOH(aq) + H_2O \rightleftarrows CH_3COO^-(aq) + H_3O^+(aq) \qquad (18.3)$$

Of the original neutral acetic acid molecules, only a small fraction ionizes. Hence, the resulting solution is a *poor conductor*. Most organic acids and bases ionize only to a small extent. Solutes that ionize only slightly and solutions that contain relatively few ions are referred to as **weak electrolytes.**

A third class of solutes includes molecular compounds that do not ionize when dissolved in water. The most minute solute particle remains a neutral molecule. Consequently, such solutions do not conduct an electric current. Solutes whose solutions are *nonconductors* are referred to as **nonelectrolytes.**

Solution conductivity can be detected by a conductivity-sensing apparatus such as the one shown in Figure 18–1. If the electrodes are immersed in a solution containing mobile ions, the ions conduct "current," and the bulb lights. If the solution contains no ions, no current will flow, and the bulb does not light. The flow or non flow of current is therefore a clear indication of the presence or absence of ions. The intensity of light produced (which is proportional to the magnitude of current) is a qualitative measure of the number of ions present.

Figure 18.1
Conductivity-sensing apparatus

In this experiment, you will observe the conductivity of several solutions. You will be asked to make classifications, based on your experimental observations, of whether the substances tested are strong electrolytes, weak electrolytes, or nonelectrolytes.

SAFETY PRECAUTIONS

Do not, at any time, handle the conductivity-testing apparatus unless specifically directed by your instructor. Never touch both electrodes at the same time; if the apparatus is plugged in, a severe electrical shock may result.

PROCEDURE

The instructor will set up a conductivity-sensing apparatus. This apparatus will be used to test the conductivity of several solutions. Classify each solution in the following list as a good conductor, a poor conductor, or a nonconductor, and record your classification in the space provided on the work page. Also, answer the corresponding questions in the Discussion section of the work page as the experiment progresses.

1. Deionized water

2. Tap water

Explain the difference between the conductivities of deionized water and tap water.

3. Solid NaCl

4. 1.0 M NaCl (salt of a strong acid and strong base). NaCl is an ionic solid.

Explain why solid NaCl did or did not conduct. Would you predict that molten NaCl (pure NaCl in liquid state) would or would not conduct? Why?

5. Glacial acetic acid, CH_3COOH or $HC_2H_3O_2$ (pure acetic acid, no water present)

6. 1.0 M $HC_2H_3O_2$

How would you explain the difference in conductivity between pure acetic acid and dissolved acetic acid?

7. 1.0 M HCl

Compare the conductivities of Solutions 6 and 7. Explain the difference between the two acids.

8. 1.0 M NaOH

9. 1.0 M NH_3(aq) (also referred to as NH_4OH)

How would you explain the difference in the observed conductivities?

10. 1.0 M $NaC_2H_3O_2$ (salt of a weak acid and strong base)

11. 1.0 M $NH_4C_2H_3O_2$ (salt of a weak acid and weak base)

12. 1.0 M NH_4Cl (salt of a strong acid and weak base)

Did you find any substantial difference between the conductivities of these solutions? Did they differ from the conductivities of 1.0 M NaCl? Did you find that it made a difference in conductivity whether the soluble salt was formed from a weak or strong acid and a strong or weak base?

13. Solid dextrose (sugar), $C_6H_{12}O_6$

14. 1.0 M dextrose (sugar)

Based on the conductivity of the sugar solution, would you classify it as a molecular solution or an ionic solution? Does it behave like pure water did, or is it similar to a salt solution? Explain your answer. Would you predict that molten sugar is a good conductor, weak conductor, or nonconductor?

15. $BaSO_4$ in water

16. Equal volumes of 1.0 M $BaCl_2$ and 1.0 M Na_2SO_4

Was there any visible evidence of a chemical reaction? Write the conventional equation as well as the net ionic equation for the reaction. Based on your observations for Species 3 and 15, explain the conductivity behavior of the mixture.

Experiment 18

Advance Study Assignment

1. What is the difference between a strong electrolyte and a weak electrolyte?

2. What do we mean by ionic compound and molecular compound?

3. How does a conductivity-sensing device, such as the one used in this experiment, indicate whether a solution contains ionic or molecular species?

Experiment 18

Work Page

Number	Substance	Conductor (Good, poor, nonconductor)	Electrolyte		
			Strong	Weak	Non
1.	Deionized water				
2.	Tap water				
3.	Solid NaCl				
4.	1.0 M NaCl				
5.	Glacial $HC_2H_3O_2$				
6.	1.0 M $HC_2H_3O_2$				
7.	1.0 M HCl				
8.	1.0 M NaOH				
9.	1.0 M NH_3				
10.	1.0 M $NaC_2H_3O_2$				
11.	1.0 M $NH_4C_2H_3O_2$				
12.	1.0 M NH_4Cl				
13.	Solid dextrose				
14.	1.0 M dextrose				
15.	$BaSO_4$ in water				
16.	1.0 M $BaCl_2$ and 1.0 M Na_2SO_4				

Discussion

1; 2: Explain the difference between the conductivities of deionized water and tap water.

3; 4: Did solid NaCl conduct? Yes _____ No _____. Explain why or why not.

Would you predict that molten NaCl would _____ or would not _____ conduct? Explain your reasoning.

5; 6: Explain the reason for the difference in conductivities.

6; 7: The acid having higher conductivity is _____. Give an explanation for the difference.

8; 9: Compare the two bases and explain the difference in conductivities.

Experiment 18

Work Page

10–12: Were the conductivities different? Yes _____ No _____. Give reasons for your observations.

13; 14: Sugar dissolved in water gives an ionic _____ a molecular _____ solution,

which resembles water _____ a salt solution _____. Molten sugar would be a

good _____ weak _____ non _____ conductor.

15; 16: Evidence of a chemical reaction: _____.

Conventional equation: _____.

Net ionic equation: _____.

Explanation for conductivity behavior:

Experiment 18

Report Sheet

Number	Substance	Conductor (Good, poor, nonconductor)	Electrolyte		
			Strong	Weak	Non
1.	Deionized water				
2.	Tap water				
3.	Solid NaCl				
4.	1.0 M NaCl				
5.	Glacial $HC_2H_3O_2$				
6.	1.0 M $HC_2H_3O_2$				
7.	1.0 M HCl				
8.	1.0 M NaOH				
9.	1.0 M NH_3				
10.	1.0 M $NaC_2H_3O_2$				
11.	1.0 M $NH_4C_2H_3O_2$				
12.	1.0 M NH_4Cl				
13.	Solid dextrose				
14.	1.0 M dextrose				
15.	$BaSO_4$ in water				
16.	1.0 M $BaCl_2$ and 1.0 M Na_2SO_4				

Discussion

1; 2: Explain the difference between the conductivities of deionized water and tap water.

3; 4: Did solid NaCl conduct? Yes _____ No _____. Explain why or why not.

Would you predict that molten NaCl would _____ or would not _____ conduct? Explain your reasoning.

5; 6: Explain the reason for the difference in conductivities.

6; 7: The acid having higher conductivity is _____. Give an explanation for the difference.

8; 9: Compare the two bases and explain the difference in conductivities.

Experiment 18

Report Sheet

10–12: Were the conductivities different? Yes _____ No _____. Give reasons for your observations.

13; 14: Sugar dissolved in water gives an ionic _____ a molecular _____ solution,

which resembles water _____ a salt solution _____. Molten sugar would be a

good _____ weak _____ non _____ conductor.

15; 16: Evidence of a chemical reaction: _____.

Conventional equation: _____.

Net ionic equation: _____.

Explanation for conductivity behavior:

Net Ionic Equations: A Study Assignment

Performance Goals

19–1 Distinguish between a conventional equation, an ionic equation, and a net ionic equation.

19–2 Identify spectators in an ionic equation.

19–3 Given information from which you may write the conventional equation for one of the following types of reactions in aqueous solution, write the net ionic equation:

a) Ion combination reactions that produce a precipitate;

b) Ion combination reactions that produce a molecular product;

c) Reactions that yield a gas;

d) Oxidation–reduction reactions that may be described by a "single replacement" equation.

CHEMICAL OVERVIEW

The conventional equation by which a chemical change is described is adequate for most purposes, but for reactions of ionic compounds in aqueous (water) solution, the typical molecular equation has serious shortcomings. Specifically, a conventional equation may indicate formulas of reactants and products that are not present, and omit completely the formulas of the ions that are the real reactants and products. If the substances in the conventional equation that are actually present as dissociated ions are written in the form of their ions, the result is an **ionic equation.** Frequently the same ions appear on both sides of an ionic equation. Though they are present at the scene of a chemical reaction, these ions experience no chemical change themselves. They are called **spectator ions,** or simply **spectators.** Eliminating the spectators from an ionic equation leaves the **net ionic equation,** an equation that includes only the substances that actually participate in the reaction.

To illustrate these equations, consider the reaction between solutions of lead(II) nitrate and sodium iodide. When the two clear solutions are combined, insoluble lead(II) iodide precipitates. The conventional equation is

$$Pb(NO_3)_2(aq) + 2\,NaI(aq) \rightarrow PbI_2(s) + 2\,NaNO_3 \qquad (19.1)$$

Three of the compounds in the equation are soluble ionic salts that dissociate into ions when dissolved. In other words, no real solute particles having the formulas $Pb(NO_3)_2$, NaI, or $NaNO_3$ are actually present in the reaction vessel. The species actually present are the ions resulting from those species. These ions, both their identity and number, can be found by writing the equation for the dissociation of that number of solute species appearing in the conventional equation:

$$Pb(NO_3)_2(aq) \rightarrow Pb^{2+}(aq) + 2\,NO_3^{-}(aq) \qquad (19.2)$$

$$2\,NaI(aq) \rightarrow 2\,Na^{+}(aq) + 2\,I^{-}(aq) \qquad (19.3)$$

$$2\,NaNO_3(aq) \rightarrow 2\,Na^{+}(aq) + 2\,NO_3^{-}(aq) \qquad (19.4)$$

Replacing the compound formulas of dissolved substances in the conventional equation with the ions really present gives the ionic equation:

$$Pb^{2+}(aq) + 2\,NO_3^{-}(aq) + 2\,Na^{+}(aq) + 2\,I^{-}(aq) \rightarrow PbI_2(s) \qquad (19.5)$$
$$+ 2\,Na^{+}(aq) + 2\,NO_3^{-}(aq)$$

The ionic equation contains two sodium ions and two nitrate ions on both sides. These are the spectators that do not undergo any change, even though they are in the solution where chemical change is occurring. Eliminating the spectators from the ionic equation gives the net ionic equation:

$$Pb^{2+}(aq) + 2\,I^{-}(aq) \rightarrow PbI_2(s) \qquad (19.6)$$

The net ionic equation isolates the two reactants that actually change chemically and the single new substance produced.

The above example does more than simply illustrate a net ionic equation. It also shows how that equation is developed. There are three steps in the procedure:

1. Write the conventional equation, including designations of state or solutions, (s) for solid, (ℓ) for liquid, (g) for gas, and (aq) for dissolved species (Equation 19.1). Be sure the equation is balanced.[*]

2. Write the ionic equation, replacing those species that ionize in aqueous solution with ions actually present (Equation 19.5). Check to be sure the equation is still balanced, both in atoms and in electrical charge.

3. Write the net ionic equation by eliminating the spectators (Equation 19.6). If necessary, reduce coefficients to their lowest integral (whole number)

---

[*]If the purpose is simply to write the net ionic equation for a reaction, it is not essential that the conventional and ionic equations be balanced. Most instructors, however, require balancing all three equations. That procedure will be followed in this book; you should follow the procedure recommended by your instructor.

values. Check to be sure the equation is still balanced, both in atoms and electrical charge.

Regardless of what you may or may not have been doing up to this point in your chemistry course, it is particularly important that you include designations of state or solution in your conventional equation. These designations identify the formulas that *may* be rewritten in the form of dissociated ions in the ionic equation. *Under no circumstances do you change the formula of a solid, liquid, or gas; those substances appear in the ionic and net ionic equations in exactly the same form as they appear in the conventional equation.* It follows that you must have sufficient information in the description of the reaction to determine the proper designations for all substances. Reactant designations are determined by careful reading of the description of the reaction. Product designations may or may not be indicated. It is generally left to you to decide whether a product formed by the combination of ions is or is not soluble, and, even if it is soluble, whether it does or does not dissociate into ions.

Solubility.   The solubility of an ion combination product may be determined by referring to Table 19.1. Directions for its use appear beneath the table. Notice that the intersection of the horizontal lead(II) ion line with the vertical iodide ion column contains the letter "s," indicating that lead(II) iodide is an insoluble solid, written as $PbI_2(s)$ in an equation. The intersection of the horizontal sodium ion line with the vertical nitrate ion column shows "aq," indicating that sodium nitrate is soluble in water, and is written $NaNO_3(aq)$ in an equation. It is worth noting that all entries in the sodium ion line and the nitrate ion column are designated "aq," showing that all sodium salts and all nitrate salts included in the table are soluble. These and several other useful generalizations are listed in the following "solubility rules":

1. All ammonium and alkali metal salts are soluble.

2. All nitrates are soluble.

3. All chlorides and bromides are soluble except those of Ag^+, Hg_2^{2+}, and Pb^{2+} ions.

4. All sulfates are soluble except those of Ca^{2+}, Sr^{2+}, Ba^{2+}, Hg_2^{2+}, Hg^{2+}, Pb^{2+}, and Ag^+ ions.

5. All carbonates are insoluble except ammonium carbonate and the carbonates of the alkali metals.

6. All hydroxides are insoluble except those of the alkali metals, $Ba(OH)_2$ and $Sr(OH)_2$.

7. All sulfides are insoluble except those of the alkali metals and ammonium sulfide.

Molecular Products—Weak Acids.   Sometimes the hydrogen ion from a strong acid will combine with the anion of a second reactant to form a molecular product. If the anion is the hydroxide ion, the molecular product is water, $H_2O(\ell)$, which appears in that form in all conventional and ionic equations. Another molecular product forms when the hydrogen ion from a strong acid combines with the anion of a weak acid. These acids ionize

Table 19.1 *Solubilities of Ionic Compounds**

Cations \ Anions	Acetate	Bromide	Carbonate	Chlorate	Chloride	Fluoride	Hydrogen Carbonate	Hydroxide	Iodide	Nitrate	Nitrite	Phosphate	Sulfate	Sulfide	Sulfite
Aluminum	S	aq		aq	aq	S		S	—	aq		S	aq	—	
Ammonium	aq	aq	aq	aq	aq	aq	aq	—	aq	aq	aq	aq	aq	aq	aq
Barium	—	aq	S	aq	aq	S		aq	aq	aq	aq	S	S	—	S
Calcium	aq	aq	S	aq	aq	S		S	aq	aq	aq	S	S	S	S
Cobalt(II)	aq	aq	S	aq	aq	—		S	aq	aq		S	aq	S	S
Copper(II)	aq	—			aq	aq		S		aq		S	aq	S	
Iron(II)	aq	aq	S		aq	S		S	aq	aq		S	aq	S	S
Iron(III)	—	aq			aq	S		S	—	aq		S	aq	—	
ead(II)	aq	S	S	aq	S	S		S	S	aq	aq	S	S	S	S
ithium	aq	aq	aq	aq	aq	aq	aq	aq	aq	aq	aq	S	aq	aq	
agnesium	aq	aq	S	aq	aq	S		S	aq	aq	aq	S	aq	—	aq
ickel		aq	S	aq	aq	aq		S	aq	aq		S	aq	S	S
otassium	aq	aq	aq	aq	aq	aq	aq	aq	aq	aq	aq	aq	aq	aq	aq
Silver	S	S	S	aq	S	aq		—	S	aq	S	S	S	S	S
Sodium	aq	aq	aq	aq	aq	aq	aq	aq	aq	aq	aq	aq	aq	aq	aq
inc	aq	aq	S	aq	aq	aq		S	aq	aq		S	aq	S	S

*To determine the solubility of an ionic compound, locate the intersection of the horizontal row for the cation and the vertical column for the anion. An "aq" in that box indicates that the compound is soluble in water to a molarity of 0.1 or more at 20°C. An "s" indicates that the compound is not soluble to that concentration, but remains in the solid state or precipitates if the ions are combined. A blank space indicates lack of data, and a dash (—) identifies an unstable substance.

only slightly, and when writing ionic equations are considered to exist in molecular form. Hydrofluoric acid, HF, for example, happens to be an extremely corrosive acid, but it is also a weak acid because the individual solute particles in its water solution are predominantly un-ionized HF molecules. Its proper designation is therefore HF(aq) in ionic equations as well as conventional equations. In other words, *a weak acid is not separated into ions in writing an ionic equation, but is written in molecular form.*

All of this makes it necessary for you to distinguish between weak acids and strong acids. Fortunately, there are only a few strong acids—few enough for you to memorize them easily. They are the three most common acids, sulfuric (H_2SO_4), nitric (HNO_3), and hydrochloric (HCl), and three other acids containing halogens, hydrobromic (HBr), hydroiodic (HI), and perchloric ($HClO_4$). If any of these six acids appears in a conventional equation, it must be broken down to ions in an ionic equation. All other acids are written in molecular form in both conventional and ionic equations.[*]

Unstable Ion Combination Products. There are three common substances that are unstable in water solution and must be written in both the conventional and ionic equations in the form of their final decomposition products. These are carbonic acid, H_2CO_3, sulfurous acid, H_2SO_3, and what is commonly called "ammonium hydroxide," whose formula is usually written NH_4OH. Carbonic and sulfurous acids break down into water plus carbon dioxide or sulfur dioxide, respectively. Carbon dioxide is relatively insoluble and bubbles out of the solution as a gas, $CO_2(g)$. Sulfur dioxide is relatively soluble and is written as $SO_2(aq)$ in the equation. "Ammonium hydroxide," NH_4OH, never really forms from NH_4^+ and OH^- ions; instead, an equilibrium is reached, $NH_4^+(aq) + OH^-(aq) \rightleftarrows NH_3(aq) + H_2O(\ell)$, in which the products are by far the predominant species and are written in molecular form in all equations.

To summarize:

What breaks down (forms ions)?

1. Strong acids;
2. Strong bases;
3. Soluble salts.

What does not break down?

1. Weak acids (remember, if it is not one of the strong acids, it is considered a weak acid!);
2. Weak bases;
3. Gases;
4. Solids;
5. Elements.

EXAMPLES

The following examples are in the form of a program in which you learn by answering a series of questions. Obtain an opaque shield—a piece of cardboard or a folded piece of paper you cannot see through—that is wide enough to cover this page. In each example, place the shield on the book page so it covers everything beneath the first dotted line that runs across the page. Read to that point, and write in the space provided whatever is asked. Then lower the shield to the next dotted line. The material exposed will begin with the correct response to the question you have just answered. Compare this answer to yours, looking back to correct any misunderstanding if the two are different. When you fully understand the first step, read to the next dotted line and proceed as before.

---

[*]Some fairly common acids, notably phosphoric, oxalic, and chloric acids, are "borderline" acids that might be considered either strong or weak, depending on some arbitrary classification standard. We will avoid these acids in this exercise.

Example 1

Write the net ionic equation for the reaction between solutions of barium bromide and ammonium sulfate.

Begin with the conventional equation. You will have to predict the products of the reaction, which is not difficult if you recall that the reaction between two ionic substances is frequently described by a "double displacement" type of equation, in which the positive ion of one reactant combines with the negative ion of the other, and vice versa—a "change partners" type of equation. Be sure to include state or solution designations, using Table 19.1 to determine the proper designations for the products.

....

1a. $BaBr_2(aq) + (NH_4)_2SO_4(aq) \rightarrow BaSO_4(s) + 2\ NH_4Br(aq)$

The (aq) designations for the reactants are indicated by the word "solutions" in the statement of the question. The intersection of the barium ion line and sulfate column line in Table 19.1 has an "s" in it, indicating that barium sulfate is a solid (precipitate), and therefore its designation in the equation is (s). The ammonium ion and bromide ion intersection shows "aq," indicating that the compound is soluble in water.

Now write the ionic equation by replacing those species that are in solution, designated (aq), by the ionic particles actually present in the solution. Remember, only those substances designated (aq) may be changed in this step. Be sure your equation is balanced.

....

1b. $Ba^{2+}(aq) + 2\ Br^-(aq) + 2\ NH_4{}^+(aq) + SO_4{}^{2-}(aq)$
$\rightarrow BaSO_4(s) + 2\ Br^-(aq) + 2\ NH_4{}^+(aq)$

The $BaSO_4(s)$, a solid, remains in the same form in the ionic equation as it appeared in the conventional equation. The other substances, all ionic compounds, dissociate into their ions.

From here all you must do is cross out the spectators and you have the net ionic equation.

....

1c. $Ba^{2+}(aq) + SO_4{}^{2-}(aq) \rightarrow BaSO_4(s)$

A final check on the equation shows one barium ion on each side, one sulfate ion on each side, and a total charge of zero on both sides. The importance of the charge balance will appear shortly.

The next example is similar, but slightly different....

Example 2

Write all equations that result in the net ionic equation for the reaction between solutions of silver nitrate and magnesium chloride.

You know the procedure for all three steps. Proceed, but be careful on the last step. . . .

....

2a. $2\,AgNO_3(aq) + MgCl_2(aq) \rightarrow 2\,AgCl(s) + Mg(NO_3)_2(aq)$

$2\,Ag^+(aq) + 2\,NO_3^-(aq) + Mg^{2+}(aq) + 2\,Cl^-(aq) \rightarrow$
$$2\,AgCl(s) + Mg^{2+}(aq) + 2\,NO_3^-(aq)$$

$2\,Ag^+(aq) + 2\,Cl^-(aq) \rightarrow$
$$2\,AgCl(s),\ \text{or simply}\ Ag^+(aq) + Cl^-(aq) \rightarrow AgCl(s)$$

The net ionic equation that results from the elimination of the magnesium and nitrate ion spectators has coefficients of 2 for all species. Step 3 in the procedure for writing net ionic equations (page 246) states that all coefficients should be expressed in their lowest terms. Dividing the first equation in the last line above by 2 yields the second equation.

At some point you should become aware of the possible combination of solutions that leads to the conditions in the next example.

Example 3

Write the net ionic equation for the reaction, if any, that occurs when a solution of sodium nitrate is poured into a solution of ammonium sulfate.

Proceed, but when you come to something you haven't seen before, think about it. . . .

....

3a. $2\,NaNO_3(aq) + (NH_4)_2SO_4(aq) \rightarrow Na_2SO_4(aq) + 2\,NH_4NO_3(aq)$

$2\,Na^+(aq) + 2\,NO_3^-(aq) + 2\,NH_4^+(aq) + SO_4^{2-}(aq) \rightarrow$
$$2\,Na^+(aq) + SO_4^{2-}(aq) + 2\,NH_4^+(aq) + 2\,NO_3^-(aq)$$

There is no net ionic equation because every species in the ionic equation is a spectator. In other words, there is no reaction, no chemical change, no reactant destroyed, no new substance formed. The final mixture contains the four ions that were initially divided between two solutions. When you encounter a question such as this, simply state "No reaction."

A neutralization reaction is one between an acid and a base, yielding water and a salt as its products. The next example gives the net ionic equation that describes all neutralizations between solutions of a strong acid and a strong base.

Example 4

Write the net ionic equation for the reaction between aqueous potassium hydroxide and sulfuric acid.

The procedure is the same as before, but there will be a small difference that shows up in your state designations. Complete the equation.

....

4a. $H_2SO_4(aq) + 2\ KOH(aq) \rightarrow 2\ H_2O(\ell) + K_2SO_4(aq)$

$2\ H^+(aq) + SO_4{}^{2-}(aq) + 2\ K^+(aq) + 2\ OH^-(aq)$
$$\rightarrow 2\ H_2O(\ell) + SO_4{}^{2-}(aq) + 2\ K^+(aq)$$

$2\ H^+(aq) + 2\ OH^-(aq) \rightarrow 2\ H_2O(\ell)$, or simply $H^+(aq) + OH^-(aq)$
$$\rightarrow H_2O(\ell)$$

The difference in state designation is that water is a liquid, designated (ℓ), rather than a solid precipitate as in Examples 1 and 2.

A common error frequently appears in writing the net ionic equation for a reaction involving sulfuric acid, as in Example 4. In writing the ionic equation, the acid is incorrectly broken into the sulfate ion and an $H_2{}^+(aq)$ ion, which does not exist. The formula for an aqueous hydrogen ion is $H^+(aq)$, whether it comes from H_2SO_4, HCl, or any other acid. The fact that elemental hydrogen is diatomic, H_2, has nothing to do with the aqueous hydrogen ion.

Example 5

Write the net ionic equation for the reaction between hydrochloric acid and solid barium hydroxide.

The reaction is very similar to Example 4, but you must read carefully....

....

5a. $2\ HCl(aq) + Ba(OH)_2(s) \rightarrow 2\ H_2O(\ell) + BaCl_2(aq)$

$2\ H^+(aq) + 2\ Cl^-(aq) + Ba(OH)_2(s) \rightarrow 2\ H_2O(\ell) + Ba^{2+}(aq) + 2Cl^-(aq)$

$2\ H^+(aq) + Ba(OH)_2(s) \rightarrow 2\ H_2O(\ell) + Ba^{2+}(aq)$

In this example it is stated that *solid* barium hydroxide is the reactant, even though barium hydroxide is sufficiently soluble to be classified (aq) in Table 19.1. Watch for words in the statement of a reaction that indicate the state of a reactant, and be sure to use that state in the equation.

The next neutralization reaction yields a slightly different net ionic equation.

Example 6

Write the net ionic equation for the reaction between hydrochloric acid and a solution of sodium fluoride.

Proceed as before, but remember the discussion of weak acids on page 247.

....

6a. $HCl(aq) + NaF(aq) \rightarrow HF(aq) + NaCl(aq)$

$H^+(aq) + Cl^-(aq) + Na^+(aq) + F^-(aq) \rightarrow HF(aq) + Na^+(aq) + Cl^-(aq)$

$H^+(aq) + F^-(aq) \rightarrow HF(aq)$

In this case the product HF(aq) is a weak acid—recognized as such because it is not one of the six strong acids—and therefore does not separate into ions.

The base used to neutralize an acid is not necessarily a hydroxide, as the next example shows.

Example 7

Write the net ionic equation for the reaction between hydrochloric acid and a solution of sodium hydrogen carbonate.

Again proceed, but this time remember the discussion of unstable ion combination products on page 249.

....

7a. $HCl(aq) + NaHCO_3(aq) \rightarrow H_2O(\ell) + CO_2(g) + NaCl(aq)$

$H^+(aq) + Cl^-(aq) + Na^+(aq) + HCO_3^-(aq)$
$\qquad\qquad \rightarrow H_2O(\ell) + CO_2(g) + Na^+(aq) + Cl^-(aq)$

$H^+(aq) + HCO_3^-(aq) \rightarrow H_2O(\ell) + CO_2(g)$

This time the ion combination product is unstable carbonic acid, H_2CO_3, which decomposes into water and carbon dioxide.

The reaction between an active metal and an acid releases hydrogen gas and leaves a solution of a salt behind. This kind of oxidation–reduction reaction is described by a conventional equation that is sometimes called a single replacement equation, in which one element appears to replace another.

Example 8

Write the net ionic equation for the reaction between zinc and sulfuric acid.

....

8a. $Zn(s) + H_2SO_4 (aq) \rightarrow H_2(g) + ZnSO_4(aq)$

$Zn(s) + 2H^+(aq) + SO_4^{2-} (aq) \rightarrow H_2(g) + Zn^{2+}(aq) + SO_4^{2-}(aq)$

$Zn(s) + 2\,H^+(aq) \rightarrow H_2(g) + Zn^{2+}(aq)$

In the conventional equation it looks as if zinc replaces hydrogen to form a compound with the sulfate ion. The net ionic equation shows the true character of the reaction, in which two electrons move from a zinc atom to two hydrogen ions. It is this transfer of electrons that classifies the reaction as an oxidation–reduction reaction.

Example 8 differs from all of the earlier examples in that each side of the equation has a net electrical charge, instead of having all charges cancel out to zero. The requirement is that the charges be *balanced*, but not necessarily at zero. If you balance the conventional equation and correctly represent the number of ions in the ionic equation that come from the conventional equation, the charges in the net ionic equation will already be balanced. They should be checked, however, and not assumed. If the charge is not balanced, something is wrong someplace in the three steps. It is possible to write a net ionic equation that is balanced in atoms but not in charge. The final example is such a reaction.

Example 9

Write the net ionic equation for the reaction between copper and a solution of silver nitrate.

The conventional equation for this oxidation–reduction reaction between copper and aqueous silver nitrate is a single replacement equation in which the copper(II) ion is produced. This hint should enable you to write the three equations.

....

9a. $Cu(s) + 2\,AgNO_3(aq) \rightarrow 2\,Ag(s) + Cu(NO_3)_2(aq)$

$Cu(s) + 2\,Ag^+(aq) + 2\,NO_3^-(aq) \rightarrow 2\,Ag(s) + Cu^{2+}(aq) + 2\,NO_3^-(aq)$

$Cu(s) + 2\,Ag^+(aq) \rightarrow 2\,Ag(s) + Cu^{2+}(aq)$

An error in the first or second step could have produced $Cu(s) + Ag^+(aq) \rightarrow Ag(s) + Cu^{2+}(aq)$ as a net ionic equation that is balanced in atoms. The net charge on the left side of the equation is 1+, however, and the charge on the right side is 2+. The equation is therefore not balanced, even though both copper and silver check out.

DEMONSTRATION (OPTIONAL)

Observe the following process carried out by the instructor, and write the net ionic equation for the reaction. Also, explain the reason for the changes during the titration.

Procedure. Pour 30 to 35 mL of clear, saturated barium hydroxide solution (carefully decanted from a bottle containing excess solute on the bottom) into a 400-mL beaker. Add about 100 mL of deionized water and 3 drops of phenolphthalein indicator. Immerse the electrodes of a conductivity-sensing device (see Figure 18.1) into the solution.

In a solution, the current-carrying species are ions; hence, good conduction indicates a great number of ions. Conversely, poor conduction shows that a relatively small number of ions are present. Observe whether the barium hydroxide solution in the beaker is a good or a poor conductor of electricity.

From a buret, slowly add a 1 M sulfuric acid solution to the contents of the beaker. Stir the solution continuously and observe any changes that occur.

Experiment 19

Advance Study Assignment

1. An ionic equation has $2H^+(aq)$ on the left side and $H_2(g)$ on the right. Are these species spectators? Explain.

2. Which of the following substances are *not* ionized in an aqueous solution? Explain.

 a) HI; b) KBr; c) H_2SO_4; d) HNO_2; e) AgCl; f) $Ca(NO_3)_2$

3. Consider the statement: After balancing a net ionic equation, the sum of the charges on each side of the equation should be equal. Is this true or false? Give an example.

Experiment 19

Work Page

Instructions: For each reaction described below, write the conventional equation, ionic equation, and net ionic equation. Include designations of state or solution in each equation.

1. Lead(II) nitrate and magnesium sulfate solutions are combined.

2. Barium metal is dropped into hydrochloric acid.

3. Potassium hydroxide solution reacts with nitric acid.

4. Zinc chloride solution is poured into a solution of ammonium carbonate.

5. Ammonium chloride and potassium hydroxide solutions are combined.

6. Magnesium chloride solution is mixed with nickel nitrate solution.

7. Sulfuric acid reacts with a solution of magnesium acetate, $Mg(C_2H_3O_2)_2$.

8. Zinc reacts with a solution of nickel sulfate.

9. Cobalt(II) sulfate and lithium sulfide solutions are combined.

10. Potassium nitrite solution is added to sulfuric acid.

11. Solid nickel carbonate is dropped into nitric acid.

12. Iron(III) chloride and aluminum sulfate solutions are mixed.

13. Hydrochloric acid is poured into a solution of lithium carbonate.

Experiment 19

Work Page

14. Hydrogen is released when sodium reacts with water.

15. Hydroiodic acid reacts with a solution of ammonium sulfite.

16. Solid copper(II) hydroxide is "dissolved" by hydrochloric acid.

17. Sodium hydroxide solution is poured into a solution of cobalt(II) chloride.

18. Calcium metal reacts with a solution of iron(II) bromide.

19. Hydrochloric acid reacts with a solution of sodium acetate, $NaC_2H_3O_2$.

20. Aluminum bromide and ammonium fluoride solutions are combined.

Demonstration Questions (Optional)

1. What were the color and conductivity of the initial barium hydroxide and phenolphthalein solution?

2. What changes occurred during the titration?

3. Why did the pink color disappear?

4. Why did the light go out?

5. Why did the light come back on?

6. Write the conventional and the net ionic equations for the reaction.

Experiment 19

Report Sheet

Instructions: For each reaction described below, write the conventional equation, ionic equation, and net ionic equation. Include designations of state or solution in each equation.

1. Lead(II) nitrate and magnesium sulfate solutions are combined.

2. Barium metal is dropped into hydrochloric acid.

3. Potassium hydroxide solution reacts with nitric acid.

4. Zinc chloride solution is poured into a solution of ammonium carbonate.

5. Ammonium chloride and potassium hydroxide solutions are combined.

6. Magnesium chloride solution is mixed with nickel nitrate solution.

7. Sulfuric acid reacts with a solution of magnesium acetate, $Mg(C_2H_3O_2)_2$.

8. Zinc reacts with a solution of nickel sulfate.

9. Cobalt(II) sulfate and lithium sulfide solutions are combined.

10. Potassium nitrite solution is added to sulfuric acid.

11. Solid nickel carbonate is dropped into nitric acid.

12. Iron(III) chloride and aluminum sulfate solutions are mixed.

13. Hydrochloric acid is poured into a solution of lithium carbonate.

Experiment 19

Report Sheet

14. Hydrogen is released when sodium reacts with water.

15. Hydroiodic acid reacts with a solution of ammonium sulfite.

16. Solid copper(II) hydroxide is "dissolved" by hydrochloric acid.

17. Sodium hydroxide solution is poured into a solution of cobalt(II) chloride.

18. Calcium metal reacts with a solution of iron(II) bromide.

19. Hydrochloric acid reacts with a solution of sodium acetate, $NaC_2H_3O_2$.

20. Aluminum bromide and ammonium fluoride solutions are combined.

Demonstration Questions (Optional)

1. What were the color and conductivity of the initial barium hydroxide and phenolphthalein solution?

2. What changes occurred during the titration?

3. Why did the pink color disappear?

4. Why did the light go out?

5. Why did the light come back on?

6. Write the conventional and the net ionic equations for the reaction.

Titration of Acids and Bases: An Introduction

Experiments 20 and 21 are really two parts of a single experiment having a single ultimate objective, namely, the determination of the concentration of an acid solution. This special overview has been written to point out the division of the procedure into two parts, and to furnish you with a basis for understanding how these parts are related. In brief, Experiment 21 uses a sodium hydroxide solution of *known* concentration to find the unknown concentration of the acid. The sodium hydroxide solution used in Experiment 21 is prepared, and its concentration is determined, in Experiment 20. It is therefore essential that the solution you prepare in Experiment 20 be kept for use in Experiment 21.

MOLARITY: A CONCENTRATION UNIT

In these experiments you will use **molarity** as your concentration unit. By definition, molarity is the number of moles of solute per liter of solution. So you may see clearly how molarity may be found from experimental data—so you may see what information you require to calculate molarity and how you use that information—the idea is presented here side by side with a more familiar parallel:

Rate of travel is expressed by speed, the units of which are miles per hour, or miles/hour.

Solution concentration is expressed by molarity, the units of which are moles per liter, or moles/liter.

If you were to hike 12.0 miles in 4.00 hours, you would find your average speed by dividing 12.0 miles by 4.00 hours:

If you were to dissolve 12.0 moles of solute in 4.00 liters of solution, you would find your molarity by dividing 12.0 moles by 4.00 liters:

$$\frac{12.0 \text{ miles}}{4.00 \text{ hours}} = 3.00 \text{ miles/hour}$$

$$\frac{12.0 \text{ miles}}{4.00 \text{ liters}} = 3.00 \text{ moles/liter}$$

Your calculation method matches the units in which speed is expressed: if you divide the number of miles traveled in a given trip by the hours taken for the same trip, the result is average speed.

Your calculation method matches the units in which molarity is expressed: if you divide the number of moles in a given sample of solution by the number of liters in the same sample, the result is the molarity of the solution.

The two essential items you require for calculation of speed are *miles traveled* and *hours in trip*.	The two essential items you require for calculation of molarity are *moles of solute* and *liters of solution*.

The last sentence in the right column sets your objective for both experiments: to find the number of moles of solute in a sample of solution and the volume of that sample. Dividing one by the other yields the required molarity.

TITRATION: A LABORATORY PROCESS

Titration is the controlled addition of a solution into a reaction vessel from a buret. By means of titration, the volume of solution used may be determined quite precisely. The titration process is used in many analytical determinations, including those involving acid–base reactions.

An **indicator** is a substance used to signal when the titration arrives at the point at which the reactants are stoichiometrically (or chemically) equal, as defined by the reaction equation. For example, in an acid–base titration between sodium hydroxide and hydrochloric acid,

$$NaOH(aq) + HCl(aq) \longrightarrow H_2O(\ell) + NaCl(aq)$$

the indicator should tell when the numbers of moles of NaOH and HCl are exactly equal, matching the 1:1 ratio in the equation. For the reaction

$$2\,NaOH(aq) + H_2SO_4(aq) \longrightarrow 2\,H_2O(\ell) + Na_2SO_4(aq)$$

the indicator should tell when the number of moles of NaOH is exactly twice the number of moles of H_2SO_4, this time reflecting the 2:1 molar ratio between the reactants. This point of chemical equality is called the **equivalence point** of the titration. Acid–base indicators send their signal by changing color at or very near the equivalence point of the titration.

A **standard solution** is a solution with a precisely determined concentration. Initially the concentration of a standard solution is determined from a weighed quantity of a **primary standard,** a highly purified reference chemical. A standard solution may be prepared in either of two ways:

1. A primary standard is carefully weighed, dissolved, and diluted accurately to a known volume. Its concentration can be calculated from the data.

2. A solution is made to an approximate concentration and then standardized by titrating an accurately weighed quantity of a primary standard.

Once a solution has been standardized in one reaction, it may be used as a standard solution in subsequent reactions. Thus the standard solution prepared in Experiment 20 will be used in the reaction of Experiment 21 to determine the concentration of an unknown acid.

With this background, we now proceed to Experiments 20 and 21 individually.

Titration of Acids and Bases–I

Performance Goals

20–1 Given the volume of a solution of known molarity, and the volume to which it is diluted with water, calculate the molarity of the diluted solution.

20–2 Given the approximate molarity and volume of an acid or base solution to be used in a titration, calculate the number of grams of a known solid base or acid required for the reaction.

20–3 Given the volume of a base or acid solution that reacts with a weighed quantity of a primary standard acid or base, calculate the molarity of the base or acid solution.

20–4 Perform acid–base titrations reproducibly.

CHEMICAL OVERVIEW

In this experiment you will prepare a standard solution of sodium hydroxide to be used in Experiment 21. Solid sodium hydroxide has the property of absorbing moisture from the air. It is therefore not possible to weigh sodium hydroxide accurately, which makes it unsuitable as a primary standard. Consequently, you will use the second of the two methods for preparing a standard solution listed on page 268. Your primary standard will be oxalic acid dihydrate, $H_2C_2O_4 \cdot 2\,H_2O$. The reaction between the acid and base is

$$2\,NaOH(aq) + H_2C_2O_4(aq) \longrightarrow 2\,H_2O(\ell) + Na_2C_2O_4(aq) \qquad (20.1)$$

Sodium hydroxide will be made available in the laboratory in the form of a solution that is approximately one molar (1 M) in concentration. Note that this is an *approximate* concentration, expressed in *one* significant figure. No calculation based on that concentration can be considered reliable. You will be instructed to dilute a specified quantity of that solution to a larger volume with water, and then to calculate the *approximate* concentration of the diluted solution (see Performance Goal 20-1). This is one of two preliminary calculations in this experiment, and it appears as Question 1 in the Advance Study Assignment. If you turn in your Advance Study Assignment at the beginning of the laboratory period, be sure to keep a copy of your calculation for use while performing the experiment. The diluted NaOH solution will be used in the titration.

Next it will be necessary for you to calculate the quantity of solid oxalic acid dihydrate, $H_2C_2O_4 \cdot 2\ H_2O$, that will react with approximately 15 mL of the diluted NaOH solution. This is a solution stoichiometry problem in which the first step is to find the number of moles of NaOH in 15 mL of solution at the approximate concentration just determined. The volume of a solution times its molarity yields the number of moles:

$$\text{Volume (liter)} \times \text{molarity}\left(\frac{\text{moles}}{\text{liter}}\right) = \text{moles} \qquad (20.2)$$

The balance of the problem is set up and solved in the usual stoichiometry pattern. This is the second of the two preliminary calculations, and it corresponds to Performance Goal 20–2. This calculation appears as Question 2 in the Advance Study Assignment. Again, be sure to keep a copy for use while performing the experiment.

After carefully weighing out three samples of solid oxalic acid dihydrate, you will dissolve it and perform the titration described by Equation 20.1. From the mass of oxalic acid you will be able to determine the number of moles of acid present. From the equation you will determine the number of moles of NaOH required in the neutralization. You will then know both the volume of the NaOH solution and the number of moles of NaOH it contains. These are the "two essential items you require for (the) calculation of molarity" (see Introduction, p. 267): dividing moles by liters yields molarity. (See Performance Goal 20–3.)

The last of the performance goals for this experiment calls for you to perform titrations reproducibly. To meet this requirement you must come up with sodium hydroxide concentrations that are "the same" in separate titrations. This calls for establishing a standard of "sameness." You will be instructed to conduct three titrations as a minimum. If two of these yield molarities that are within 0.007 M of each other, they will be accepted as satisfying the reproducibility requirement. If you do not reach this result, additional titrations will be required.

So far nothing has been mentioned about the *accuracy* of your work. Indeed, within Experiment 20 there is no way to judge accuracy, because each student will have his/her own sodium hydroxide solution, which will have a concentration slightly different from that of his/her neighbor. In Experiment 21, however, you will use your standard solution to determine the concentration of an acid that is unknown to you, *but known to your instructor*. At this point an accurate result will be required. It should be apparent that your result in Experiment 21 cannot be accurate unless the concentration of the solution prepared in Experiment 20 has been determined accurately. As a consequence, if your accuracy in Experiment 21 does not meet the standard established, *it may be necessary for you to repeat Experiment 20 in order to correct previously undetected errors made there*. With this in mind, you are strongly urged to retain a complete record of *all* data, of *all* volumes in the titrations, even if you think they might be incorrect. It is surprising how often "incorrect" data turn out to be just what is needed by the time a long experiment is completed.

SAMPLE CALCULATIONS

The following examples illustrate the calculations involved in Performance Goals 20–1 and 20–3:

Example 1

25.0 mL of a 12.0 M solution is diluted to 500 mL. Calculate the molarity of the dilute solution.

The number of moles of solute is the same in both the initial solution and the diluted solution; only water is added. This number of moles is

$$0.0250 \text{ L} \times \frac{12.0 \text{ moles}}{\text{L}} = 0.300 \text{ mole}$$

In the diluted solution the 0.300 mole of solute is dissolved in 500 mL, or 0.500 L. The concentration is therefore

$$\frac{0.300 \text{ mole}}{0.500 \text{ L}} = 0.600 \text{ mole/L}$$

Example 2

Calculate the molarity of an NaOH solution if a sample of oxalic acid weighing 1.235 g requires 42.5 mL of the base for neutralization.

First, determine the number of moles of oxalic acid present. The formula of solid oxalic acid is $H_2C_2O_4 \cdot 2\,H_2O$. Notice that, although the water of hydration is not shown in Equation 20.1, the acid is weighed as a solid, which includes the water. Accordingly, calculations must be based on the proper molar mass. Thus,

$$1.235 \text{ g } H_2C_2O_4 \cdot 2H_2O \times \frac{1 \text{ mole } H_2C_2O_4}{126 \text{ g } H_2C_2O_4 \cdot 2H_2O} = 0.00980 \text{ mole } H_2C_2O_4$$

Second, determine the number of moles of NaOH required to react with 0.00980 mole of $H_2C_2O_4$, according to the equation.

$$0.00980 \text{ mole } H_2C_2O_4 \times \frac{2 \text{ moles NaOH}}{1 \text{ mole } H_2C_2O_4} = 0.0196 \text{ mole NaOH}$$

Third, if 0.0196 mole of NaOH is present in 42.5 mL of solution, find the concentration in moles per *liter*.

$$\frac{0.0196 \text{ mole NaOH}}{0.0425 \text{ L}} = 0.461 \text{ M NaOH}$$

As a single dimensional analysis setup, this calculation would appear as:

$$1.235 \text{ g } H_2C_2O_4 \cdot 2H_2O \times \frac{1 \text{ mole } H_2C_2O_4}{126 \text{ g } H_2C_2O_4 \cdot 2H_2O} \times \frac{2 \text{ moles NaOH}}{1 \text{ mole } H_2C_2O_4} \times \frac{1}{0.0425 \text{ L}} = 0.461 \text{M NaOH}$$

SAFETY PRECAUTIONS AND DISPOSAL METHODS

If some of the sodium hydroxide solution, either concentrated or dilute, comes into contact with your skin, it will have a slippery feeling, somewhat like soap. This is produced because the solution is slowly dissolving a layer of your skin. For obvious reasons, the process should not be allowed to continue. If you encounter that feeling at any time during the experiment, take time out to wash your hands thoroughly until the slippery feeling is gone.

Both the acid and base used in this experiment are corrosive and harmful over prolonged exposure. Avoid all unnecessary contact, and keep them off your clothing too. Both solutions are harmful to the eyes; be sure to wear goggles when working with all chemicals, either solid or in solution, during this experiment. The goggle requirement extends to cleaning-up operations, as well.

After you have finished the titrations, SAVE your NaOH solution for Experiment 21. The content of the Erlenmeyer flasks can be poured down the drain.

PROCEDURE

NOTE: Record all mass measurements in grams to the nearest 0.001 g. Record all volume measurements from the buret in milliliters to the nearest 0.1 mL.

Preparation of NaOH Solution

A. Using a graduated cylinder, transfer about 100 mL of 1 M NaOH to a large beaker (600 mL or larger). With continuous stirring, dilute with deionized water until the total volume is about 500 mL.

NOTE: Thorough mixing is essential at this point. If you determine the "concentration" of an unmixed solution, you determine the concentration of only that part of the solution that you use. If you then use another part of the solution, with a different concentration, you will have no accuracy in your second application.

B. Transfer your solution to a stoppered or capped storage bottle. (Always keep your standard solution covered, because any evaporation loss or CO_2 absorption will change its concentration.) Label the bottle with your name, so it does not become lost among the bottles of your laboratory neighbors.

C. Calculate the approximate molarity of your solution. (This is Question 1 in the Advance Study Assignment.)

Preparation of Oxalic Acid Solutions

A. From the approximate molarity of the diluted NaOH solution, calculate the mass of oxalic acid dihydrate, $H_2C_2O_4 \cdot 2 H_2O$, needed to neutralize 15 mL of the base. Don't forget the water of hydration in the solid acid, because it will be present in what you weigh out. Have your instructor approve your calculation before proceeding. (This is Question 2 in the Advance Study Assignment.)

B. Make identifying marks on three 250-mL Erlenmeyer flasks.

NOTE: The next step is extremely critical. Your purpose is to transfer into each of the above Erlenmeyer flasks an amount of oxalic acid dihydrate that is approximately equal to what was calculated in Step 2A, but whose actual mass is known to the closest milligram. The care with which this step is performed will determine both the accuracy and precision with which your sodium hydroxide solution is standardized, and the accuracy of your result in Experiment 21. Be particularly careful that no oxalic acid is spilled. If that happens, begin again.

C. Label three **dry** Erlenmeyer flasks. Using a milligram balance, weigh the first flask to the milligram and record the mass of the empty flask. From a beaker, using a spatula, transfer some oxalic acid crystals into the flask, being very careful not to spill any. The mass of the acid in each flask should be approximately equal to the amount you calculated in Step 2A. The amount of acid transferred can be calculated by taking the mass of the flask plus acid and subtracting the mass of the empty flask. Repeat the procedure with the other flasks.

If you do not have dry flasks available, use a weighing paper instead. Weigh the empty paper first, then add the oxalic acid. Be very careful when you transfer the solid into the labeled Erlenmeyer flasks not to lose any crystals.

D. Add about 75 mL of deionized water to each flask. Swirl or use a magnetic stirrer to *completely* dissolve the crystals.

Titration of Oxalic Acid with NaOH

A. Rinse your buret with deionized water. Now, rinse the buret with about 10 mL of *your dilute* NaOH solution. Make sure the tip of the buret has also been rinsed. Drain and discard the rinse and repeat the rinsing with a second portion of NaOH. Fill the buret with NaOH, using a funnel. Drain the solution into the calibrated portion of the buret, letting the liquid flow through the tip. Make sure there is no bubble present under the stopcock. Set the bottom of the meniscus on a whole line—never start between lines! Record the buret reading as the initial buret reading on the work page.

B. Add 3 to 5 drops of phenolphthalein to each oxalic acid solution.

C. If you have a magnetic stirring apparatus available, carefully add a stirring bar to the flask containing the completely dissolved acid. Position the flask such that the tip of the buret is inside the neck of the flask. Start stirring at a moderate speed. If you do not have such apparatus available, you will have to swirl the flask manually throughout the titration (see Figure 20.1).

Start the addition of NaOH. At the beginning of the titration you may add the base in larger portions, slowing down as the time required for the pink color to disappear gets longer. It is essential that you mix (or swirl) the contents of the flask throughout the titration to ensure complete mixing of the solutions. The end of the titration is reached when the pink color persists for 30 seconds. Record the final buret reading on the work page.

Figure 20.1
Titration from a buret into a flask.

D. Adjust the volume so that the bottom of the meniscus is on a whole line. Do not start from the final volume of the previous run if this volume was between lines! Be sure there is sufficient volume of NaOH left to complete the second titration. Remember, the bottom portion of the buret is not calibrated! If this is not the case, fill up the buret (using a funnel), adjust the liquid level, record the initial reading, and proceed with the second titration as described above.

E. Calculate the results of the first three runs immediately. If the range of the molarity values is greater than 0.007, repeat the process with two more samples. *Keep all titration data.*

If you satisfy the reproducibility requirement, proceed to Experiment 21. If that experiment is to be done at another time, store your NaOH solution in an upright position (see Step 1B). Clean all glassware thoroughly and rinse it with deionized water before putting it away or returning it to the stockroom.

Experiment 20

Advance Study Assignment

1. Read Part 1 for the preparation of your NaOH solution. Calculate the approximate molarity of the solution (Step 1C). Use three significant figures.

2. Using your result from Question 1, calculate the mass of oxalic acid dihydrate, $H_2C_2O_4 \cdot 2 H_2O$, required to react with 15 mL of your NaOH solution (Step 2A in the procedure). Round off your answer to three significant figures. It is this result that must be approved by your instructor before you go beyond Step 2A in the procedure.

3. If 0.628 g of $H_2C_2O_4 \cdot 2 H_2O$ requires 24.7 mL of NaOH(aq) in a titration, calculate the molarity of the solution.

4. Why must you save your NaOH solution at the end of Experiment 20?

5. If your fingers feel slippery at any time during this experiment, what does that indicate and what should you do about it?

Experiment 20

Work Page

TABLE OF DATA AND RESULTS
(Beneath the table show the full calculation setup for at least one valid titration run)

Sample	1	2	3	4	5	6	7	8
Mass of flask + acid (g)								
Mass of flask (g)								
Mass of oxalic acid (g)								
Initial buret reading (mL)								
Final buret reading (mL)								
Volume of NaOH (mL)								
Molarity of NaOH								

Average molarity ———

Calculation setup for at least one valid titration run:

Experiment 20

Report Sheet

TABLE OF DATA AND RESULTS
(Beneath the table show the full calculation setup for at least one valid titration run)

Sample	1	2	3	4	5	6	7	8
Mass of flask + acid (g)								
Mass of flask (g)								
Mass of oxalic acid (g)								
Initial buret reading (mL)								
Final buret reading (mL)								
Volume of NaOH (mL)								
Molarity of NaOH								

Average molarity ———

Calculation setup for at least one valid titration run:

Titration of Acids and Bases—II

Performance Goals

21–1 Determine the concentration of an acid by titrating a known volume with a standardized base.

CHEMICAL OVERVIEW

In this experiment you will use the standardized NaOH from Experiment 20 to determine the molar concentration of either hydrochloric acid (Option 1) or vinegar (Option 2), the latter being an acetic acid solution. The reactions for the two options are:

$$\text{HCl(aq)} + \text{NaOH(aq)} \rightarrow \text{H}_2\text{O}(\ell) + \text{NaCl(aq)} \qquad \text{(Option 1)} \qquad (21.1)$$

$$\text{HC}_2\text{H}_3\text{O}_2\text{(aq)} + \text{NaOH(aq)} \rightarrow \text{H}_2\text{O}(\ell) + \text{NaC}_2\text{H}_3\text{O}_2\text{(aq)} \qquad \text{(Option 2)} \qquad (21.2)$$

You will titrate a carefully measured volume of the acid with the NaOH solution. The product of the volume of NaOH times its molarity is the number of moles of base in the reaction. This is converted to moles of acid from the stoichiometry of the reaction. The number of moles of acid divided by the volume containing that number of moles (the volume of the acid sample) yields the molarity of the acid. These ideas were presented more fully in the general overview to Experiments 20 and 21.

To be acceptable, your reported molarity for the unknown must be within 0.015 of the molarity determined previously. Your results must also show good precision, that is, good reproducibility. In this experiment all sample sizes are the same. Therefore, all titration volumes should be alike. The requirement for precision will be satisfied if you have duplicate titrations within 0.2 mL (about 6 drops) of each other.

If your results are unsatisfactory from the standpoint of either precision or accuracy, your instructor may ask you to make repeat runs until acceptable results are reached. If precision is good but accuracy is bad, it is quite likely that the error lies in your results from Experiment 20. In this case your standardization runs against oxalic acid as a primary standard must be repeated. Keep all titration results, because they are helpful in locating errors if they do appear.

SAFETY PRECAUTIONS AND DISPOSAL METHODS

The same safety precautions identified for Experiment 20 apply to Experiment 21—plus one more. In taking your samples of unknown acid, you will use a volumetric pipet. *Always* use a pipet bulb to draw liquid chemicals into a pipet; never use mouth suction.

After you have finished your titrations and checked your results with the instructor, the dilute NaOH can be poured down the drain and rinsed with plenty of cold water or use the disposal method specified by your instructor.

PROCEDURE

NOTE: Record all volume measurements in milliliters to the nearest 0.1 mL.

A. Clean and prepare your buret for use as in Experiment 20, Step 3A. Be sure to shake the NaOH solution before you start using it, especially if the solution was stored for a period of time. This will ensure that the water condensed on the walls of your storage bottle is reabsorbed into the solution. Fill the buret with the standardized NaOH solution from Experiment 20.

B. Pipet a 10.0-mL sample of your unknown acid—hydrochloric acid (Option 1) or vinegar (Option 2)—into each of three 250-mL Erlenmeyer flasks. Add about 50 mL of deionized water and 3 to 5 drops of phenolphthalein to each flask.

C. Titrate each acid sample, using the same procedure that you used in Step 3C of Experiment 20. If the three runs do not yield at least two volumes that are within 0.2 mL of each other, run three additional samples. Record and retain all data.

D. Calculate the molarity of your unknown acid and submit it for approval if your instructor so requests. If it is not approved, complete any additional titrations, either standardizations of the NaOH or analyses of the unknown, as required.

E. On receiving approval of your final result (if required by your instructor), thoroughly clean and rinse all glassware before putting it away or returning it to the stockroom.

Experiment 21

Advance Study Assignment

1. Why must you use a pipet bulb to draw a liquid into a pipet?

2. Why must you swirl the contents of the flask during the titration?

3. Why is it important that you shake the NaOH solution before using it in this experiment?

4. Calculate the molarity of an acetic acid solution, $HC_2H_3O_2(aq)$, if a 25.0-mL sample requires 27.2 mL of 0.138 M NaOH in a titration.

Experiment 21

Work Page

TABLE OF DATA AND RESULTS
(Beneath the table show the full calculation setup for at least one valid titration run)

Unknown number _____ Average molarity of NaOH from Experiment 20 _____

Sample	1	2	3	4	5	6	7	8
Initial buret reading (mL)								
Final buret reading (mL)								
Volume of NaOH (mL)								
Molarity of HCl (Option 1)								
Molarity of acetic acid (Option 2)								

Calculation setup for at least one valid titration run:

Experiment 21

Report Sheet

TABLE OF DATA AND RESULTS
(Beneath the table show the full calculation setup for at least one valid titration run)

Unknown number _____ Average molarity of NaOH from Experiment 20 _____

Sample	1	2	3	4	5	6	7	8
Initial buret reading (mL)								
Final buret reading (mL)								
Volume of NaOH (mL)								
Molarity of HCl (Option 1)								
Molarity of acetic acid (Option 2)								

Calculation setup for at least one valid titration run:

Determination of a Chemical Equation

Performance Goals

22–1 By dilution, prepare a known molarity solution of sulfuric acid and sodium hydroxide.

22–2 Using the dilute solutions, titrate a given volume of acid with the base.

22–3 Based on the titration data, calculate the mole ratio of base to acid and write the chemical equation corresponding to this mole ratio.

CHEMICAL OVERVIEW

The concept of molarity and the method of titration are discussed in detail in the introduction to Experiments 20 and 21 on pages 267–68. Please read this material carefully before beginning this experiment.

In this experiment you will be supplied two "concentrated" solutions of known molarity, from which you will prepare, by dilution, solutions that you will use in the titration. You will then calculate the molarity of the dilute solutions (see Example 1 in the Sample Calculations).

To perform the titration, first you will carefully pipet a given volume of sulfuric acid into three Erlenmeyer flasks. Then, you will titrate each sample with the dilute sodium hydroxide. A minimum of three titrations will be carried out. Since all acid samples have the same volume, the amount of sodium hydroxide should be nearly the same. If the three runs are not within the range of 0.2 mL, additional titrations will have to be performed.

Using the volumes and molarity of the sulfuric acid and sodium hydroxide, you will calculate the number of moles contained in each. Using these values, the mole ratio can be calculated (see Example 2 in the Sample Calculations). *Based on this ratio*, you will then write the chemical equation representing the reaction that has taken place.

SAMPLE CALCULATIONS

Example 1

15.0 mL of a 3.12 M solution is diluted to 250.0 mL. Calculate the molarity of the dilute solution.

The number of moles of solute is the same in both solutions, since only water was added. This can be expressed mathematically by the equation

$$M_1V_1 = M_2V_2$$

where M_1 and V_1 are the molarity and volume of the concentrated solution, respectively, and M_2 and V_2 those of the dilute solution. Using the equation above and solving for M_2, we get

$$M_2 = \frac{M_1V_1}{V_2}; \qquad M_2 = \frac{3.12 \text{ moles}}{L} \times \frac{15.0 \text{ mL}}{250.0 \text{ mL}} = 0.187 \text{ M}$$

Note that we do not necessarily have to convert the volumes to liters, since the volume ratio will remain the same regardless of what units are used—as long as they are the same!

Example 2

In a titration, 15.0 mL of 0.152 M sulfuric acid required 31.5 mL of 0.145 M sodium hydroxide to reach the end point. Calculate the mole ratio of base to acid in this titration.

First, calculate the number of moles of acid and base, using the relationship

$$\text{moles} = M \times V; \qquad \text{moles} = (\text{moles/L}) \times L$$

Note, in this case you *must* convert the volume to liters!

$$\text{moles H}_2\text{SO}_4 = \frac{0.152 \text{ mole}}{L} \times 0.0150 \text{ L} = 0.00228 \text{ mole}$$

$$\text{moles NaOH} = \frac{0.145 \text{ mole}}{L} \times 0.0315 \text{ L} = 0.00457 \text{ mole}$$

$$\frac{\text{moles NaOH}}{\text{moles H}_2\text{SO}_4} = \frac{0.00457}{0.00228} = 2$$

This means that 2 moles of NaOH reacts with 1 mole of H_2SO_4. Your experimental values may not yield an exact whole number; round your ratios to the nearest whole number if you are within 0.15 of a whole number (e.g., 2.89 rounds to 3; 0.95 rounds to 1).

SAFETY PRECAUTIONS AND DISPOSAL METHODS

Both sodium hydroxide and sulfuric acid are very corrosive and harmful substances. If some of the sodium hydroxide solution, either concentrated or dilute, comes in contact with your skin, it will feel slippery. Wash your skin immediately with lots of cold water until the slippery feeling is gone. Sulfuric acid will burn when in contact with skin; rinse the exposed area with plenty of cold water.

Obviously, you must avoid any contact with your eyes. Be sure to wear goggles throughout the experiment and during cleanup. Also, always use a pipeting device, never use mouth suction!

Dispose of the solutions as directed by your instructor.

PROCEDURE

NOTE: Record all volume measurements in milliliters to the nearest 0.1 mL.

---

1. Preparation of NaOH Solution

A. Obtain a 15.0-mL pipet, a pipet bulb (or similar device), and a 250.0-mL volumetric flask. Rinse the pipet and the flask with deionized water. Next, rinse the pipet (only!) with about 5 mL of the "concentrated" NaOH, and discard the rinse. Pipet 15.0 mL of the NaOH into the volumetric flask. Add deionized water from a beaker until the water level is about 1 cm below the etch line on the neck of the flask. Add the remaining water drop by drop using an eye dropper until the *bottom* of the meniscus just touches the line (see Figure LP.4). Be careful, because if you overfill the flask, you will have to start over again! Place the stopper in the flask and mix the solution very thoroughly. Turning the flask upside down, at least five times, while gently shaking it will accomplish this. This step is *extremely* important, because a non-homogeneous solution will never yield proper duplication of titration values.

B. Transfer the diluted NaOH to a *dry*, capped storage bottle. Keep this bottle closed, except when pouring from it, because evaporation loss or absorption of CO_2 will change the concentration.

C. Record the molarity of the "concentrated" NaOH and calculate the concentration of the diluted solution. Record your value on the work page.

2. Preparation of H₂SO₄ Solution

A. Rinse the pipet and the volumetric flask with deionized water. Rinse the pipet (only!) with the "concentrated" sulfuric acid solution. Follow the procedure in Part 1, except use "concentrated" sulfuric acid.

B. Store the diluted acid in a *dry*, 250-mL Erlenmeyer flask.

3. Titration of H₂SO₄ with NaOH

A. Set up three 250-mL Erlenmeyer flasks. Rinse your pipet first with deionized water, then with about 5 mL of your *diluted* H_2SO_4 solution. Discard the rinse. Now, pipet 15.0-mL portions of the acid into each of the Erlenmeyer flasks. Add about 75 mL of deionized water and 3–5 drops of phenolphthalein indicator to each flask.

B. Rinse a 50-mL buret first with deionized water, then with two 10-mL portions of the *diluted* NaOH, discarding the rinse. Always use a funnel to fill the buret. Make sure the tip of the buret is also rinsed. Fill up the buret with the *diluted* NaOH, set the bottom of the meniscus on a line (not necessarily a whole milliliter), and read the volume. Record this value on your work page (this is the initial NaOH). Be sure the tip of the buret is filled and that you don't see a bubble under the stopcock. If there is one, let down some solution from the buret and set the level again.

C. If a magnetic stirring apparatus is available, place a stirring bar into the flask and position the tip of the buret inside the neck of the Erlenmeyer flask (see Figure 20.1). If you do not have magnetic stirring available, you will have to swirl the flask manually throughout the titration.

At the beginning of the titration you may add the base in larger portions, slowing down as the time for the pink color to disappear gets longer. The end of the titration is reached when the pink color persists for 30 seconds. Record the buret reading on your work page (this is the final NaOH).

D. Repeat Step C with the remaining two samples. Be sure to fill up the buret if you judge that not enough solution is available for the next titration. Adjust the level on a whole line (never start between lines!). Remember, the bottom part of the buret is not calibrated!

E. If the three titrations are not in the required range of 0.2 mL, pipet two more acid samples and repeat the titration. You should have three values within 0.2 mL of each other.

F. When you are finished, dispose of the solutions as instructed. Return the magnetic stirring bar to the container from which you obtained it.

CALCULATIONS

From the titration data and the diluted molarities, calculate the number of moles of H_2SO_4 and NaOH present in the volumes used (see Example 2 in the Sample Calculations). Then, calculate the mole ratio for each valid run, moles of NaOH/moles of H_2SO_4. Round your values to the nearest whole number. Use the average of the mole ratios to write the chemical equation for the reaction that has occurred.

Experiment 22

Advance Study Assignment

1. How many moles of H_3PO_4 are contained in 75.0 mL of 0.215 M solution?

2. A 20.0-mL sample of 0.875 M HCl solution was diluted to 150.0 mL. Calculate the molarity of the resultant solution.

3. Why do you have to rinse the pipet with the "concentrated" H_2SO_4 before using it to prepare the dilute solution?

4. Why is it advisable to keep the storage bottle containing the NaOH solution closed?

Experiment 22

Work Page

Concentration Data

Solution	Concentrated	Dilute
H_2SO_4		
NaOH		

Titration Data

Sample	1	2	3	4	5	6
Final NaOH (mL)						
Initial NaOH (mL)						
Volume NaOH (mL)						
Moles NaOH						
Volume H_2SO_4 (mL)						
Moles H_2SO_4						
Mole Ratio $\dfrac{\text{Moles NaOH}}{\text{Moles } H_2SO_4}$						

Show calculations for one valid titration run:

Write the chemical equation that corresponds to your mole ratio:

Experiment 22

Report Sheet

Concentration Data

Solution	Concentrated	Dilute
H_2SO_4		
NaOH		

Titration Data

Sample	1	2	3	4	5	6
Final NaOH (mL)						
Initial NaOH (mL)						
Volume NaOH (mL)						
Moles NaOH						
Volume H_2SO_4 (mL)						
Moles H_2SO_4						
Mole Ratio $\dfrac{\text{Moles NaOH}}{\text{Moles } H_2SO_4}$						

Show calculations for one valid titration run:

Write the chemical equation that corresponds to your mole ratio:

A Study of Reaction Rates

Performance Goal

23–1 State qualitatively and demonstrate experimentally the relationship between the rate of a chemical reaction and (a) temperature, (b) reactant concentration, and (c) the presence of a catalyst.

CHEMICAL OVERVIEW

The word *rate* implies change and time—change in some measurable quantity and the interval of time over which the change occurs. Speed in miles per hour, salary in dollars per month, and quantity in gallons per minute all express the idea of rate. The rate of a chemical reaction tells us how fast a reactant is being consumed or a product is being produced. Specifically, rate of reaction is the positive quantity that denotes how the concentration of a species in the reaction changes with time. For example, in the reaction

$$N_2(g) + 3\,H_2(g) \rightarrow 2\,NH_3(g) \tag{23.1}$$

the rate of reaction can be expressed in terms of NH_3:

$$\text{Rate} = \frac{\text{change in concentration of } NH_3}{\text{time interval}} \tag{23.2}$$

Alternatively, the rate could be expressed in terms of a reactant:

$$\text{Rate} = -\frac{\text{change in concentration of } N_2}{\text{time interval}} \tag{23.3}$$

The minus sign is necessary to make the rate a positive quantity since the concentration of N_2 is decreasing with time.

The rate of a chemical reaction can be changed by (a) varying the concentration of reactants, (b) changing the temperature, or (c) introducing a catalyst.* In this experiment, you will study the effect of concentration by varying the concentration of one reactant while holding others constant, you will examine the rates of the same reaction at different temperatures, and you will conduct two reactions that will be identical in all respects except that a catalyst will be present in one and not in the other.

---

* Catalysts are substances that drastically alter reaction rates without being used up in the reaction.

In analyzing the results of this experiment, *time* of reaction will be used as an indication of rate. Time and rate are inversely related: the higher the rate, the shorter the time. This is evident if you think of driving from one location to another: it takes less time if you drive at a higher rate (speed).

The principal reaction whose rate will be studied in this experiment is the iodine "clock reaction." It is the reaction between two solutions, solution A and solution B, that contain, among other things, an iodide and soluble starch. When the two solutions are combined, a series of reactions begins, ending with the release of elemental iodine. The appearance of iodine in the presence of starch may be detected easily by the appearance of a deep blue color. This signals the completion of the reaction.

SAFETY PRECAUTIONS AND DISPOSAL METHODS

Sulfuric acid can cause severe burns when in contact with skin. If you should spill any of the 6 M acid on yourself, *immediately* wipe it off with a dry cloth or paper towel and then rinse it off with large amounts of water. Wear safety goggles when performing this experiment. Beware that sulfuric acid generates a lot of heat when it comes in contact with water.

Solutions obtained in this experiment can be poured down the drain.

PROCEDURE

1. Comparison of Reaction Rates

Sodium Oxalate and Potassium Permanganate

A. Place about 15 mL of 0.1 M sodium oxalate, $Na_2C_2O_4$, into a large test tube and add 2 dropperfuls of 6 M sulfuric acid, H_2SO_4.

B. Place in a second test tube an amount of deionized water equal to the total volume of solution in the first test tube. This will be used as a control, or a color standard with which to compare the test tube in which the reaction is taking place.

C. Add 8 drops of 0.1 M potassium permanganate, $KMnO_4$, to the control test tube and mix by stirring. Place the test tube in a test-tube rack.

D. Recording the time to the nearest second, add 8 drops of 0.1 M potassium permanganate to the sodium oxalate solution and mix by stirring. Place the test tube in the rack, and observe the gradual color change compared to the control test tube. Record the time in seconds at which the reaction is complete, shown by the solution becoming colorless (not tan). Proceed to the next step while you are waiting.

The Iodine Clock Reaction

E. Using your graduated cylinder, pour 10 mL of solution A into a test tube. Rinse your graduated cylinder with deionized water and pour into it 10 mL of solution B.

F. Place a small beaker on a piece of white paper. Noting the time of mixing to the nearest second, pour the two solutions simultaneously into the beaker. Record the time to the nearest second when you observe the first sign of a reaction, shown by the appearance of a dark blue color.

2. Effect of Concentration

To study the effect of concentration on the reaction rate, you will keep total volume, concentration of solution A, and temperature constant, and vary only the concentration of solution B.

A. Pour 10 mL of solution A into each of four test tubes, as you did into one test tube in Step 1E. Place the test tubes into a test-tube rack. Rinse your cylinder with deionized water.

B. Label four other test tubes 1 to 4 and place them into the test-tube rack. Using a 10-mL graduated cylinder, pour into each test tube the quantities of deionized water and solution B shown in the following table:

Test Tube Number	Volume of Deionized Water (mL)	Volume of Solution B (mL)	Total Volume (mL)
1	2	8	10
2	4	6	10
3	6	4	10
4	8	2	10
Part 1E	0	10	10

The undiluted solution B used in Steps 1E and 1F is added to the table for comparison. Note that the test tube from 1E has the highest concentration of solution B, test tube number 1 the next highest, and test tube number 4 the lowest. Note also that the total volume is the same in each case.

C. As in Step 1F, at some time 0 on a watch, pour the solution from test tube number 1 and the solution A from one of the four test tubes simultaneously into a small beaker on a piece of white paper. Record the number of seconds required for the reaction. Empty the beaker and rinse it with deionized water. Repeat the procedure with the remaining test-tube combinations, recording the elapsed time in each case.

3. Effect of Temperature

A. Using a graduated cylinder, measure 10 mL of solution A into each of four test tubes. Rinse the cylinder, and then measure 10 mL of solution B into each of four other test tubes. Measure and record the temperature of solution A in one of the test tubes. (We will assume that solution B is at the same temperature, which is room temperature.) As before, empty one test tube of solution A and one test tube of solution B simultaneously into a small beaker. Record the time required for the reaction. (This is a repeat of Step 1F, except that this time you have recorded the temperature of the reactants.)

B. Place a test tube of solution A and a test tube of solution B into a 250-mL beaker containing tap water and a few chunks of ice. Immerse a thermometer into the test tube containing solution A. Be sure not to let the thermometer rest on the bottom of the test tube. (Again, we will assume that solution B, treated in the same manner as solution A, will have the same temperature.) Stir the contents of the beaker with the test tubes. When the temperature drops to about 10°C below room temperature, remove both test tubes from the ice bath and immediately pour the two solutions into a beaker, as before. Record both the temperature of the solutions and the time of the reaction.

C. Repeat the above procedure, replacing the ice-water bath with warm tap water. When the temperature of solution A has risen about 10°C above room temperature, pour the two solutions together. Obtain one more reading at about 20°C above room temperature. For each reading, record the solution temperature and the reaction time on the work page.

4. Effect of a Catalyst

Measure 10 mL of solution A into a test tube and add a drop of 0.1 M copper(II) sulfate, $CuSO_4$, catalyst. In another test tube, combine 2 mL of solution B and 8 mL of deionized water (the same procedure as in Part 2, test tube number 4). Pour the solutions simultaneously into a beaker and record the time required for the reaction.

RESULTS

Part 2. Using the graph paper provided, plot a graph of milliliters of solution B divided by total milliliters versus reaction time (use two significant figures). What relationship seems to be evident between concentration of solution B and the time needed for the blue color to appear?

Part 3. Using the graph paper provided, plot a graph of temperature versus reaction time.

Part 4. Compare the reaction times of the catalyzed reaction (Part 4) and the uncatalyzed reaction (Part 2, test tube 4) of solutions of the same concentration. How does the time required for the blue color to appear in this step compare with the time you observed for test tube number 4 in Part 2? Record your comparison on the work page.

Experiment 23

Advance Study Assignment

1. Name two variables that affect reaction rates.

2. What is a catalyst?

3. Give an everyday example of how temperature alters reaction rates.

Experiment 23

Work Page

Part 1—Comparison of Reaction Rates

	Sodium Oxalate + Potassium Permanganate	Iodine Clock Reaction
Time reaction began		
Time reaction completed		
Time of reaction (sec)		

Reaction having higher rate: Sodium oxalate + potassium permanganate _____; iodine clock reaction _____.

Part 2—Effect of Concentration on the Iodine Clock Reaction

Test Tube	Volume of Solution B / Total Volume	Time (sec)
1		
2		
3		
4		

Reaction rate varies directly _____, inversely _____ with concentration of solution B.

Part 3—Effect of Temperature on the Iodine Clock Reaction

Temperature (°C)	Time (sec)

Increasing the temperature increases _____, decreases _____ the reaction rate.

Part 4—Effect of a Catalyst on the Iodine Clock Reaction

Reaction	Time (sec)
Without catalyst (test tube number 4, Part 2)	
With catalyst (from Part 4)	

Which *reaction rate* is greater? Catalyzed _____; uncatalyzed _____.

Experiment 23

Work Page

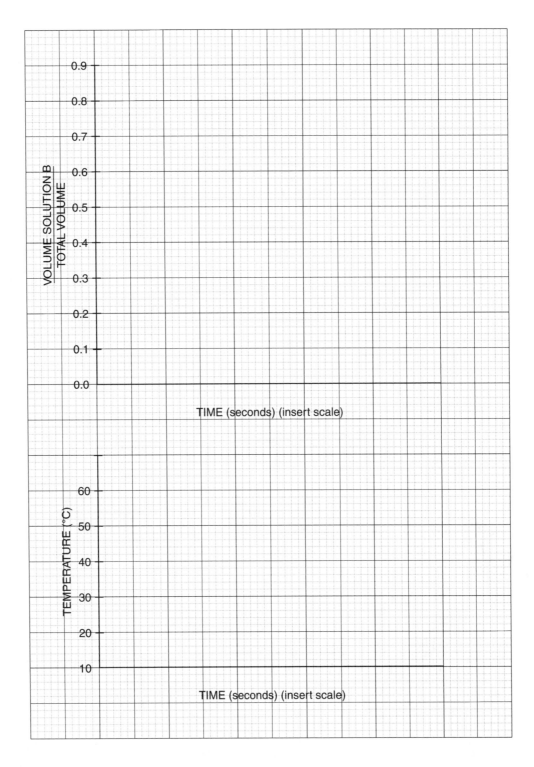

Experiment 23

Report Sheet

Part 1—Comparison of Reaction Rates

	Sodium Oxalate + Potassium Permanganate	*Iodine Clock Reaction*
Time reaction began		
Time reaction completed		
Time of reaction (sec)		

Reaction having higher rate: Sodium oxalate + potassium permanganate _____; iodine clock reaction _____.

Part 2—Effect of Concentration on the Iodine Clock Reaction

Test Tube	$\dfrac{\textit{Volume of Solution B}}{\textit{Total Volume}}$	*Time (sec)*
1		
2		
3		
4		

Reaction rate varies directly _____, inversely _____ with concentration of solution B.

Part 3—Effect of Temperature on the Iodine Clock Reaction

Temperature (°C)	Time (sec)

Increasing the temperature increases _____, decreases _____ the reaction rate.

Part 4—Effect of a Catalyst on the Iodine Clock Reaction

Reaction	Time (sec)
Without catalyst (test tube number 4, Part 2)	
With catalyst (from Part 4)	

Which *reaction rate* is greater? Catalyzed _____; uncatalyzed _____.

1. State the meaning of the term "reaction rate."
2. What is a catalyst?
3. If you wanted to alter the rate of a chemical reaction, what changes would you make in the experimental conditions?

Experiment 23

Report Sheet

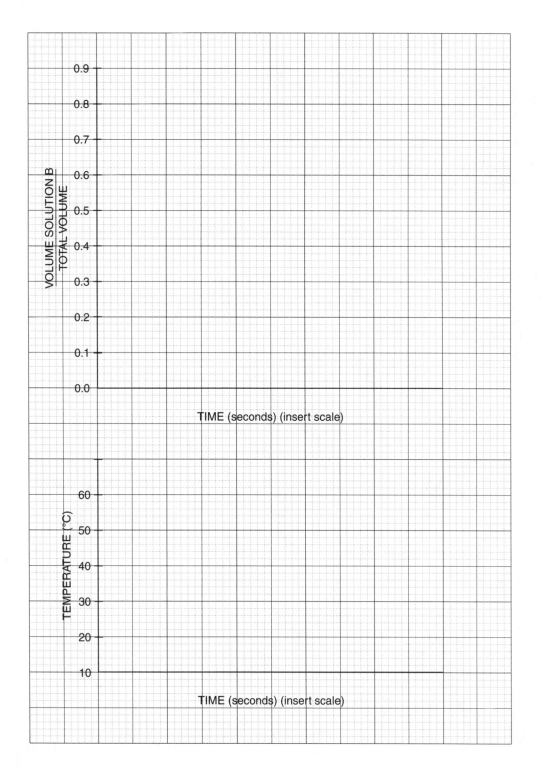

Chemical Equilibrium

Performance Goals

24–1 Given the equation for a chemical equilibrium, predict and explain, on the basis of LeChatelier's Principle, the direction of a shift in the position of an equilibrium caused by a change in the concentration of one species.

CHEMICAL OVERVIEW

Some chemical reactions *proceed to completion*, that is, until one of the reacting species is for all practical purposes completely consumed. One example of such a reaction is the precipitation of Cl^- ions from solution:

$$Ag^+(aq) + Cl^-(aq) \rightarrow AgCl(s) \tag{24.1}$$

When the Cl^- concentration is essentially zero, the reaction is complete.

Other reactions are *reversible*. This means that when the reactants are introduced into the reaction vessel, the reaction will start, but as soon as the reaction products begin to accumulate, they will react with each other to produce some of the starting species. At any time, *two* reactions are occurring, one going in the *forward* direction and one in *reverse*. For example, consider the ionization of acetic acid in water:

$$CH_3COOH(aq) + H_2O(\ell) \rightleftarrows CH_3COO^-(aq) + H_3O^+(aq) \tag{24.2}$$

While some neutral acetic acid molecules are reacting with water, some of the hydronium ions (H_3O^+) and acetate ions (CH_3COO^-) formed by the ionization are recombining to yield the undissociated acid. The two arrows indicate the simultaneous occurrence of two reactions. *When the rate of the forward reaction exactly equals the rate of the reverse reaction, the system is said to be at equilibrium,* and no more detectable change occurs. This condition does not mean that all reactions have ceased, but only that the opposing reactions proceed at the same rate.

Consider the generalized reaction

$$A + B \rightleftarrows C + D \tag{24.3}$$

When the concentration of any one of the species in this equilibrium is changed, the equilibrium is "disturbed," and a "shift" will occur, either in the forward or reverse direction. **LeChatelier's Principle** predicts the direction of such a shift by stating: *When some stress is applied to a system originally at equilibrium, the system (reaction) will shift in such a direction as to counteract the stress, until a new equilibrium is reached.*

Let us consider how we can apply the preceding principle to the reaction shown in Equation 24.3. Suppose we add more of compound A. What will happen? The outside stress is an increase in the concentration of A. The reaction will shift in a direction that will counteract this increase. That is, the reaction will shift to reduce the concentration of A. Compound A is consumed, and its concentration reduced, if the reaction shifts in the forward direction (as you read the equation from left to right). Based on the same reasoning, we can predict that if more C is added, the reverse shift will occur (consuming some of C). The displacement of an equilibrium by the addition of more of one of the species involved in the equilibrium is known as the **common ion effect.**

Evidence of a shift in equilibrium can easily be observed in the laboratory if one or more of the substances are colored or if a change in phase, such as precipitation or dissolution, accompanies the shift.

In this experiment, you will observe qualitatively the effect of changing the concentration of one or more substances in a chemical equilibrium, and then correlate your observations with LeChatelier's Principle. A chemical description of each equilibrium you will study is given in the Procedure section.

SAFETY PRECAUTIONS AND DISPOSAL METHODS

Solutions of sodium hydroxide react with your skin (or eyes), giving a slippery feeling. If this should occur, wash with plenty of water until no more slippery feeling is detected. Handle acid solutions with care, avoiding contact with skin. Stopper all reagent bottles as soon as you are through using them. Wear goggles when performing this experiment.

Solutions containing heavy metal ions (Co^{2+} and Fe^{3+}) should be disposed of in bottles provided. The rest of the solutions can be poured down the drain.

PROCEDURE

1. Cobalt(II) Ion Complexes

Cobalt(II) ions, Co^{2+}, exist in water as aquo-complexes, $Co(H_2O)_6^{2+}$, that have a pink color. Other complexes exhibit different colors; the $CoCl_4^{2-}$ complex, for example, is blue. Depending on the relative concentration of chloride ions, the equilibrium shown in the following equation may be altered to yield a solution that is more blue or more pink:

$$Co(H_2O)_6^{2+}(aq) + 4\,Cl^-(aq) \;\rightleftarrows\; CoCl_4^{2-}(aq) + 6\,H_2O(\ell) \tag{24.4}$$
$$\underset{Pink}{\phantom{Co(H_2O)_6^{2+}(aq)}} \qquad\qquad \underset{Blue}{\phantom{CoCl_4^{2-}(aq)}}$$

A. Pour about 10–15 drops of cobalt(II) chloride, $CoCl_2$, solution into each of three small test tubes. To the first one add 10 drops of concentrated hydrochloric acid, HCl. Note the change, if any.

B. To the second test tube, add a small amount of solid ammonium chloride, NH_4Cl, and shake to make a saturated solution. You should have undissolved solid on the bottom of the test tube.

C. Compare the color of the solution with that in the third test tube, which contains only the cobalt(II) chloride solution. Place test tubes 2 and 3 into a beaker containing boiling water and note the change, if any. Cool the tubes under running cold water. Tabulate and explain your observations. [*Hint:* The dissolution of NH_4Cl is endothermic.]

2. Ionization of Acetic Acid

Consider the equilibrium

$$CH_3COOH(aq) + H_2O(\ell) \rightleftarrows CH_3COO^-(aq) + H_3O^+(aq) \qquad (24.5)$$

Since no species in this reaction is colored, an auxiliary reagent is needed to help detect any shift in the equilibrium. You will use an indicator, methyl orange, for this purpose. In strongly acidic solutions (high H_3O^+ concentrations), methyl orange is red. A decrease in H_3O^+ concentration will cause a color change from red to yellow, with an intermediate color of orange.

A. Pour 10–15 drops of 0.1 M acetic acid, Ch_3COOH, into a test tube and add 1 or 2 drops of methyl orange. Place a few crystals of sodium acetate, CH_3COONa, in the solution and shake gently to dissolve them. Explain your observations.

B. Do not do experimentally, but reason out what change you would see if instead of sodium acetate, you added a few drops of 1 M NaOH. Explain.

3. Ionization of Aqueous Ammonia

When ammonia gas is dissolved in water the following equilibrium is established:

$$NH_3(aq) + H_2O(\ell) \rightleftarrows NH_4^+(aq) + OH^-(aq) \qquad (24.6)$$

This solution is sometimes also called ammonium hydroxide because of the presence of ammonium and hydroxide ions, but more properly it should be called aqueous ammonia. As with the equilibrium in Part 2, no species is colored. Therefore, we will use an indicator to signal the shifts in equilibrium. Phenolphthalein is colorless in slightly basic solutions, but it turns pink as the OH^- ion concentration increases.

A. Pour 10–15 drops 0.1 M ammonia, NH_3, into a test tube and add 1 or 2 drops of phenolphthalein. Note the color of the solution. Add solid ammonium chloride, NH_4Cl, to the solution and shake gently to dissolve the crystals. Record and explain your observations.

B. Repeat the procedure in Step A, but instead of solid ammonium chloride, add 10 drops of 1 M zinc chloride, $ZnCl_2$. Record and explain your observations. [*Hint:* $Zn(OH)_2$ is quite insoluble.]

4. The Thiocyano-Iron(III) Complex Ion

The thiocyano-iron(III) complex ion can be formed from iron(III) ions (Fe^{3+}) and thiocyanate ions (SCN^-) according to the equation

$$Fe^{3+}(aq) + SCN^-(aq) \rightleftarrows Fe(SCN)^{2+}(aq) \qquad (24.7)$$
$$\underset{Tan}{\phantom{Fe^{3+}(aq) + SCN^-}} \qquad \underset{Blood\ Red}{\phantom{Fe(SCN)^{2+}}}$$

Pour 10–15 drops of 0.1 M iron(III) nitrate, $FeCl_3$, and 10–15 drops of 0.1 M potassium thiocyanate, KSCN, into a 50-mL beaker. Dilute with 25 to 30 mL of deionized water to reduce the intensity of the deep red color. Pour 2- or 3-mL portions of this solution into each of three test tubes and proceed as follows:

A. Add about 10 drops of 1.0 M iron(III) nitrate solution to the contents of the first test tube.

B. To the second test tube, add 10 drops of 0.1 M potassium thiocyanate solution.

C. To the third test tube, add 4–5 drops of 10 percent sodium hydroxide.

Record and explain any changes you observed in Parts (A) through (C). (*Hint:* Iron(III) hydroxide is very insoluble.)

5. Saturated Sodium Chloride Equilibrium

When a saturated solution of sodium chloride is in contact with undissolved solute, the following equilibrium exists:

$$NaCl(s) \rightleftharpoons Na^+(aq) + Cl^-(aq) \qquad (24.8)$$

The solution contains the maximum number of ions the solubility of NaCl allows. Pour 10–15 drops of saturated solution into a small test tube and add a few drops of concentrated hydrochloric acid, HCl. Note the result and give an explanation for it.

Experiment 24

Advance Study Assignment

1. State how and explain why the equilibrium $2X + Y \rightleftarrows Z$ will shift if

   a. X is removed.

   b. extra Y is added.

   c. some Z is added.

2. Define, state, or describe:

   a. The common ion effect

   b. LeChatelier's Principle

   c. Are there any reactions occurring when a system is at equilibrium?

3. Consider the equilibrium $2\,CrO_4^{2-}(aq) + 2H^+(aq) \rightleftarrows Cr_2O_7^{2-}(aq) + H_2O(\ell)$. Predict the direction the equilibrium will shift upon the

   a. addition of NaOH.

   b. addition of hydrochloric acid.

4. Give reasons for your predictions in Question 3.

Experiment 24

Work Page

Part 1—Cobalt(II) Ion Complex Equilibrium

$$Co(H_2O)_6{}^{2+}(aq) + 4\,Cl^-(aq) \rightleftarrows CoCl_4{}^{2-}(aq) + 6\,H_2O(\ell)$$

1A. *HCl addition:* Color change, if any: _____

Direction of shift (foward, reverse, none): _____

Explanation:

1B. *NH$_4$Cl addition:* Colors of solutions in test tubes 2 and 3 at different temperatures:

	Test Tube 2	Test Tube 3
Room temperature		
Boiling water temperature		
After cooling		

Explanation:

Part 2—Ionization Equilibrium of Acetic Acid

$$CH_3COOH(aq) + H_2O(\ell) \rightleftarrows CH_3COO^-(aq) + H_3O^+(aq)$$

2A. *CH₃COONa addition:* Color change, if any: _____

Direction of shift (forward, reverse, none): _____

Explanation:

2B. *NaOH addition:* Predicted direction of shift (forward, reverse, none): _____

Explanation:

Part 3—Ionization Equilibrium of Aqueous Ammonia

$$NH_3(aq) + H_2O(\ell) \rightleftarrows NH_4^+(aq) + OH^-(aq)$$

3A. *NH₄Cl addition:* Color change, if any: _____

Direction of shift (forward, reverse, none): _____

Explanation:

Experiment 24

Work Page

3B. *ZnCl$_2$ addition:* Color change, if any: _____

Direction of shift (forward, reverse, none): _____

Explanation:

Part 4—Thiocyano-Iron(III) Complex Ion Equilibrium

$$Fe^{3+}(aq) + SCN^-(aq) \rightleftarrows Fe(SCN)^{2+}(aq)$$

4A. *Fe(NO$_3$)$_3$ addition:* Color change, if any: _____

Direction of shift (forward, reverse, none): _____

Explanation:

4B. *KSCN addition:* Color change, if any: _____

Direction of shift (forward, reverse, none): _____

Explanation:

4C. *NaOH addition:* Color change, if any: _____

Direction of shift (forward, reverse, none): _____

Explanation:

Part 5—Saturated Sodium Chloride Equilibrium

$$NaCl(s) \rightleftarrows Na^+(aq) + Cl^-(aq)$$

HCl addition: Change, if any: _____

Direction of shift (forward, reverse, none): _____

Explanation:

Experiment 24

Report Sheet

Part 1—Cobalt(II) Ion Complex Equilibrium

$$Co(H_2O)_6^{2+}(aq) + 4\,Cl^-(aq) \rightleftarrows CoCl_4^{2-}(aq) + 6\,H_2O(\ell)$$

1A. *HCl addition:* Color change, if any: _____

Direction of shift (forward, reverse, none): _____

Explanation:

1B. *NH₄Cl addition:* Colors of solutions in test tubes 2 and 3 at different temperatures:

	Test Tube 2	Test Tube 3
Room temperature		
Boiling water temperature		
After cooling		

Explanation:

Part 2—Ionization Equilibrium of Acetic Acid

$$CH_3COOH(aq) + H_2O(\ell) \rightleftarrows CH_3COO^-(aq) + H_3O^+(aq)$$

2A. *CH$_3$COONa addition:* Color change, if any: _____

Direction of shift (forward, reverse, none): _____

Explanation:

2B. *NaOH addition:* Predicted direction of shift (forward, reverse, none): _____

Explanation:

Part 3—Ionization Equilibrium of Aqueous Ammonia

$$NH_3(aq) + H_2O(\ell) \rightleftarrows NH_4^+(aq) + OH^-(aq)$$

3A. *NH$_4$Cl addition:* Color change, if any: _____

Direction of shift (forward, reverse, none): _____

Explanation:

Experiment 24

Report Sheet

3B. *ZnCl₂ addition:* Color change, if any: _____

Direction of shift (forward, reverse, none): _____

Explanation:

Part 4—Thiocyano-Iron(III) Complex Ion Equilibrium

$$Fe^{3+}(aq) + SCN^-(aq) \rightleftarrows Fe(SCN)^{2+}(aq)$$

4A. *Fe(NO₃)₃ addition:* Color change, if any: _____

Direction of shift (forward, reverse, none): _____

Explanation:

4B. *KSCN addition:* Color change, if any: _____

Direction of shift (forward, reverse, none): _____

Explanation:

4C. *NaOH addition:* Color change, if any: _____

Direction of shift (forward, reverse, none): _____

Explanation:

Part 5—Saturated Sodium Chloride Equilibrium

$$NaCl(s) \rightleftarrows Na^+(aq) + Cl^-(aq)$$

HCl addition: Change, if any: _____

Direction of shift (forward, reverse, none): _____

Explanation:

Measurement of pH with Indicators

Performance Goals

25–1 Prepare a set of pH indicator standards.

25–2 Measure the pH of an unknown solution by using indicators.

CHEMICAL OVERVIEW

Solutions of strong electrolytes such as strong acids and strong bases are good conductors of electricity. This indicates a high concentration of ions. In fact, strong acids and bases break into ions almost completely by either of two processes. **Dissociation** is the term used to describe the release of existing ions when an ionic compound dissolves, as in

$$NaOH(s) \rightarrow Na^+(aq) + OH^-(aq) \qquad (25.1)$$

Ionization is the process whereby ions are formed when a covalent compound reacts with water, as in

$$HCl(g) + H_2O(\ell) \rightarrow Cl^-(aq) + H_3O^+(aq) \qquad (25.2)$$

Even though the terms *ionization* and *dissociation* do not mean exactly the same thing, they are closely related and are often used interchangeably.

By contrast, solutions of weak electrolytes, such as weak acids and weak bases, are poor conductors of electricity. Because current is carried by mobile ions, this indicates a low concentration of ions. We therefore conclude that weak acids and bases are only partially ionized in water solutions. When acetic acid ionizes by reaction with water, equilibrium is reached:

$$CH_3COOH(aq) + H_2O(\ell) \rightleftarrows CH_3COO^-(aq) + H_3O^+(aq) \qquad (25.3)$$

At equilibrium, acetic acid is only about 1 percent ionized, compared with HCl, which is nearly 100 percent ionized, as shown by Equation 25.2. Relatively few acetate (CH_3COO^-) and hydronium (H_3O^+) ions are present at equilibrium, but unionized acetic acid molecules (CH_3COOH) are in abundance.

The *acidity* of an aqueous solution is a measure of the concentration of the hydrogen (H^+) or hydronium (H_3O^+) ion.

NOTE: The hydronium ion may be considered a hydrated hydrogen ion, $H^+ \cdot H_2O$. The H^+ ion is easier to work with and will be used hereafter. It should be understood, however, that this ion is hydrated in aqueous solution and does not exist as a simple H^+.

A convenient way to express the low acidity of weak acids is to use the **pH** scale. The pH of a solution is mathematically related to the hydrogen ion concentration by the equation

$$pH = -\log[H^+] = \frac{1}{\log[H^+]} \qquad (25.4)$$

where $[H^+]$ is the concentration of the hydrogen ion in moles per liter. By the mathematics of this equation, pH is the negative of the exponent of 10 that expresses the hydrogen ion concentration. For example, if $pH = 5$, then $[H^+] = 10^{-5}$; and a solution whose $pH = 8$ has a hydrogen ion concentration of 10^{-8}, or $[H^+] = 10^{-8}$.

Water ionizes into hydrogen and hydroxide ions:

$$H_2O(\ell) \rightleftarrows H^+(aq) + OH^-(aq) \qquad (25.5)$$

At 25°C, the ion product of water—the hydrogen ion concentration multiplied by the hydroxide ion concentration—is equal to 1.0×10^{-14}, or

$$[H^+]\,[OH^-] = 1.0 \times 10^{-14} \qquad (25.6)$$

If the ionization of water is the only source of these ions, it follows that they must be equal in concentration:

$$[H^+] = [OH^-] = 1.0 \times 10^{-7} \qquad (25.7)$$

A solution in which the hydrogen ion concentration is equal to the hydroxide ion concentration is said to be **neutral.** The pH of a neutral solution is, by calculation, 7. If the hydrogen ion concentration is greater than the hydroxide ion concentration, the solution is said to be **acidic.** In an acidic solution, the pH is less than 7. This is because the negative exponent of 10 decreases as the hydrogen ion concentration increases. $[H^+] = 10^{-5}$ ($pH = 5$) is a larger hydrogen ion concentration than $[H^+] = 10^{-7}$ ($pH = 7$). Conversely, in **basic** solutions, the concentration of the hydroxide ion exceeds the concentration of the hydrogen ion, and the pH will be greater than 7. Note at this point that as the hydrogen ion concentration goes down, the acidity decreases and the pH increases. The pH scale is illustrated in Figure 25.1.

Indicators are organic substances that impart to a solution a color that depends on its pH. Ordinarily, the color will change gradually over a range of about two pH units. In this experiment, you will use two different indicators in a set of solutions of known pH. By comparing colors, you then estimate the pH of an unknown solution.

Figure 25.1
pH scale

SAMPLE CALCULATIONS

Example 1

Calculate the pH of a 0.001 M HNO_3 solution.

Because nitric acid is a strong acid, we assume that it is completely ionized. Therefore,

$$[H^+] = 0.001, \text{ or } 10^{-3} \text{ mole/L} \quad pH = 3$$

Example 2

Calculate the pH of a solution that contains 0.01 mole of hydrogen ion in 100 mL.

First, calculate $[H^+]$, always expressed in moles/liter:

$$\frac{0.01 \text{ mole } H^+}{0.100 \text{ L}} = 0.1 \text{ mole } H^+/L = 10^{-1} \text{ mole/L}$$

It follows that the pH is 1.

Example 3

Calculate the pH of a solution that is obtained when 25.0 mL of 2.0 M HNO_3 are diluted to 500 mL. Remember, nitric acid is assumed to be completely ionized. First, calculate the number of moles of H^+ present in 25.0 mL, or 0.0250 liter, of concentrated acid.

$$0.0250 \text{ L} \times \frac{2.0 \text{ moles } H^+}{L} = 0.050 \text{ mole } H^+$$

This many moles of hydrogen ions are present in the final volume, 500 mL. Therefore, the $[H^+]$ is

$$[H^+] = \frac{0.050 \text{ mole } H^+}{0.500 \text{ L}} = 0.10 \text{ mole/L} = 1.0 \times 1.0^{-1} \text{ mole/L}$$

The pH of the diluted solution is 1.00.

Alternately, you can carry out a dilution calculation in a single step by multiplying the original concentration by a *dilution factor*, the original volume divided by the final volume:

$$[H^+] = \underset{\substack{\text{Original} \\ \text{concentration}}}{\frac{2.0 \text{ moles } H^+}{L(\text{conc})}} \times \underset{\substack{\text{Dilution} \\ \text{factor}}}{\frac{0.0250 \text{ L (conc)}}{0.500 \text{ L (dil)}}} = \underset{\substack{\text{Final} \\ \text{concentration}}}{0.10 \text{ mole } H^+/L \text{ (dil)}}$$

PROCEDURE

1. Water Preparation

Because carbon dioxide from the air dissolves in water, yielding an acidic solution, we must remove all dissolved carbon dioxide from the water used in this experiment. Place 350 to 400 mL of deionized water in a beaker and heat it to boiling. Continue boiling for approximately 10 minutes, cover the vessel with a large watch glass, and allow it to cool to room temperature.

2. Preparation of Standard Solutions

A. While the deionized water is being prepared, wash and label six test tubes having a capacity of greater than 10 milliliters. Label them 1 to 6. When the deionized water is at room temperature, prepare a set of solutions as described in Steps B to F. Figure 25.2 is a schematic diagram of the dilution procedure.

B. As accurately as possible, measure 5.0 mL of 1.0 M HCl into a 50-mL graduated cylinder. Your 5.0-mL measurement will be more accurate if you perform it in a separate 10-mL cylinder and then transfer the solution to the larger cylinder. Dilute the 5.0-mL HCl solution to 50.0 mL, again very accurately, with the treated deionized water. Transfer the contents to a small *dry* beaker and stir thoroughly with a glass rod. Pour 10.0 mL of this solution into test tube number 1.

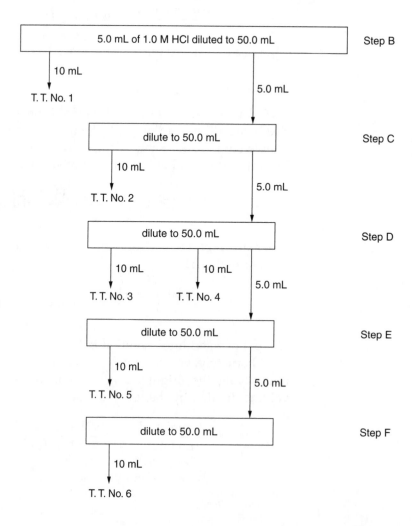

Figure 25.2
Dilution scheme

C. Carefully and accurately measure 5.0 mL of the solution prepared in Step B into a dry (or deionized-water-rinsed) 50-mL graduated cylinder and again dilute carefully to 50.0 mL with boiled deionized water. Transfer to a clean, *dry* beaker, stir, and pour 10.0 mL into test tube number 2.

D. Repeat the procedure, using 5.0 mL from Step C, diluting to 50.0 mL. This time, pour 10.0-mL samples into *two* test tubes, number 3 and number 4.

E. Dilute 5.0 mL of solution from Step D to 50.0 mL in the same fashion and pour 10.0 mL into test tube number 5.

F. Finally, dilute 5.0 mL from Step E to 50.0 mL and pour 10.0 mL into test tube number 6.

G. If you have not already done so, calculate the hydrogen ion concentration and, assuming complete ionization, the pH of each of the solutions you prepared. Enter the results on the work page.

H. To each of the test tubes numbered 1 through 3, add 2 drops of thymol blue indicator and mix well with a glass rod. *Be sure the rod is clean and dry before placing it in each solution.* Note the color in each test tube and record your observations on the work page. From these observations, estimate the pH range over which thymol blue changes color.

I. To test tubes 4 through 6, add 2 drops of methyl orange indicator and mix in the same fashion. Again record the observed color in each test tube, and estimate the pH range over which methyl orange changes color.

NOTE: If the methyl orange indicator solution is too concentrated, you will see the orange color of the indicator solution itself, regardless of what the pH is. Add only as much indicator solution as is necessary to give a clearly distinguishable color in the various test solutions, or dilute the indicator with deionized water and then add it to the test solutions.

---

3. pH of an Unknown Solution

Obtain one or more unknowns from your instructor. Pour 10.0 mL of the first unknown into each of two test tubes. To the first test tube, add 2 drops of thymol blue indicator; to the second, add 2 drops of methyl orange (see note above). Mix thoroughly. Compare the colors of these solutions to your "standard" solutions. Basing your conclusions on the colors you observe, estimate and report the pH of the unknown solution. Repeat for each unknown.

Experiment 25

Advance Study Assignment

1. If a solution has a pH of 9.5, is it acidic or basic?

2. Calculate the pH of a 0.0001 M HCl solution. If you had an acetic acid solution of the same concentration, would its pH be higher or lower? Explain.

3. Steps 1B to 1F in the Procedure section describe a series of dilutions, beginning with 5.0 mL of 1.0 M HCl. Step 1G asks you to calculate the hydrogen ion concentration and pH for each diluted solution. Perform these calculations now and enter the results in the table below.

Test Tube	$[H^+]$	pH	Test Tube	$[H^+]$	pH
1			4		
2			5		
3			6		

Experiment 25

Work Page

Test Tube Number	H$^+$ Concentration (mole/L)	pH	Indicator	Color Observed
1				
2				
3				
4				
5				
6				
Unknown No._____	XXX			
Unknown No._____	XXX			
Unknown No._____	XXX			

Estimated pH range of color transition:

a. Thymol blue: _____

b. Methyl orange: _____

Experiment 25

Report Sheet

Test Tube Number	H^+ Concentration (mole/L)	pH	Indicator	Color Observed
1				
2				
3				
4				
5				
6				
Unknown No._____	XXX			
Unknown No._____	XXX			
Unknown No._____	XXX			

Estimated pH range of color transition:

   a. Thymol blue: _____

   b. Methyl orange: _____

Introduction to Oxidation–Reduction Reactions

Performance Goal

26–1 Determine experimentally the relative strengths of a selected group of oxidizing agents.

CHEMICAL OVERVIEW

Oxidation is defined as the process in which a loss of electrons occurs; **reduction** is a gain of electrons. From a broader viewpoint, in oxidation the oxidation number of an element increases (becomes more positive, as $+3 \rightarrow +5$, or $-3 \rightarrow -1$); whereas in reduction, the oxidation number decreases (becomes more negative, as $0 \rightarrow -1$, or $+7 \rightarrow +2$).

When a metal combines chemically with a halogen to form an ionic compound, an oxidation-reduction (redox) reaction occurs. Electrons are lost by the metal and gained by the halogen. Redox reactions may be thought of as electron transfer reactions, much as acid–base reactions may be viewed as proton transfer reactions. Each redox reaction may be considered the sum of two "half-reactions" or half-cell reactions:

$M \rightarrow M^{2+} + 2\ e^-$ Oxidation of a metal to M^{2+} ion by losing 2 electrons,

and

$X_2 + 2\ e^- \rightarrow 2\ X^-$ Reduction of a halogen to X^- by gaining 1 electron per atom.

Addition produces the net ionic redox equation:

$$\begin{array}{r} M \rightarrow M^{2+} + 2\ e^- \\ X_2 + 2\ e^- \rightarrow 2\ X^- \\ \hline M + X_2 \rightarrow M^{2+} + 2\ X^- \end{array}$$

Observe that the number of electrons lost by the metal exactly equals the number of electrons gained by the halogen. This balance is essential in any redox reaction; there can never be a deficiency or excess of electrons. Just as an ordinary chemical equation has to be balanced, so does a redox equation. Balancing is achieved by adjusting one or both half-reactions in order to equate the number of electrons lost and gained.

Example 1

$$A \rightarrow A^{2+} + 2\ e^- \quad \text{Oxidation}$$
$$B + e^- \rightarrow B^- \qquad\quad \text{Reduction}$$

Multiply the reduction half-reaction by 2 and add:

$$A \rightarrow A^{2+} + 2\ e^- \quad \text{Oxidation}$$
$$\underline{2\ B + 2\ e^- \rightarrow 2\ B^-} \qquad\quad \text{Reduction}$$
$$A + 2\ B \rightarrow A^{2+} + 2\ B^- \quad \text{Net ionic redox equation}$$

Example 2

$$A \rightarrow A^{3+} + 3\ e^- \quad \text{Oxidation}$$
$$B + 2\ e^- \rightarrow B^{2-} \qquad \text{Reduction}$$

Multiply the oxidation reaction by 2 and the reduction reaction by 3 to equate the electrons lost in oxidation to the electrons gained in reduction:

$$2\ A \rightarrow 2\ A^{3+} + 6\ e^- \quad \text{Oxidation}$$
$$\underline{3\ B + 6\ e^- \rightarrow 3\ B^{2-}} \qquad\quad \text{Reduction}$$
$$2\ A + 3\ B \rightarrow 2\ A^{3+} + 3\ B^{2-} \quad \text{Net ionic redox equation}$$

In each of the preceding examples, the first element loses electrons to the second element; that is, the first element provides the electrons that reduce the second. Thus, the first element is referred to as a **reducing agent.** By accepting electrons, the second element causes the oxidation of the first element. Hence it is called an **oxidizing agent.** A summary of these terms is presented below:

If a species	*the species undergoes*	*and is called the*
Gains electrons	Reduction	Oxidizing agent (oxidizer)
Loses electrons	Oxidation	Reducing agent (reducer)

Just as acids vary in their strength (the ease with which they release protons), so reducers vary in their strength (the ease with which they release electrons). Similarly, oxidizers have different tendencies to capture electrons, just as bases vary in their attraction for protons. A strong oxidizer (oxidizing agent) has a great affinity for electrons.

In this experiment, you will investigate the relative ease with which certain metals and halides release electrons and will thereby build a partial qualitative chart of oxidizer strengths.

NOTE: Since metals and halides release electrons, they are the reducing agents, while metal ions and elemental halogens, Cl_2, Br_2, I_2, are the oxidizing agents.

Evidence of a metal–metal ion reaction is visible when a drop of solution containing the ion is placed on the metal. To detect a reaction between a halogen and a dissolved halide, however, a nonaqueous solvent that is immiscible with water must be used.

Halogens dissolved in trichloroethane have characteristic colors. Chlorine, Cl_2, in trichloroethane is colorless; bromine ranges from tan in dilute solutions to deep red or maroon when concentrated; and iodine is pale pink to deep purple, depending on concentration. You will use trichloroethane as a solvent to determine which halogen is present after a halogen and a halide ion have been combined. The trichloroethane is not involved in any chemical change; it serves simply as a solvent.

To interpret your observations correctly, consider the two examples below:

1. Suppose that the reaction between I_2 and Br^- goes to completion.

$$I_2 + 2\ Br^- \rightarrow 2\ I^- + Br_2$$

Because in the reaction Br_2 is produced, the *bottom* layer will be yellow/brown (depending on the concentration).

2. Suppose the reaction in the preceding example did not occur. Because there is no reaction, the *bottom* layer will be purple, showing that I_2 is still present. Note that simply seeing a color does not mean that a reaction has taken place!

SAFETY PRECAUTIONS AND DISPOSAL METHODS

Bromine in elemental form and concentrated iodine solution cause severe burns on contact with the skin. Also, bromine and chlorine waters release vapors that are extremely harmful when inhaled. These liquids should be handled only in a fume hood—and with utmost care. Trichloroethane vapors are harmful and should not be inhaled.

Do not pour solutions containing trichloroethane down the drain. Dispose of these solutions in a stoppered bottle.

PROCEDURE

1. Metal–Metal Ion Reactions

A. Obtain one strip each of copper, zinc, and lead. Clean one side of each strip with emery paper. Lay the strips side by side on a paper towel on the desk, cleaned surface up.

B. Place on each strip 1 drop of each solution shown in Table 26.1. Record on your work page the combinations of metals and metal ions that showed evidence of a chemical change and those that did not. Wait about 5 minutes before you decide that no reaction has occurred. Include the silver ion, Ag^+, although the metal itself is not used

Table 26.1 *Solutions for Testing Metals*

Metal	Test Solutions
Copper, Cu	Zn^{2+}, Pb^{2+}, Ag^+
Zinc, Zn	Cu^{2+}, Pb^{2+}, Ag^+
Lead, Pb	Cu^{2+}, Zn^{2+}, Ag^+

because of its high cost. Assume that silver metal does not react with Zn^{2+}, Pb^{2+}, or Cu^{2+}.

C. On completing this part of the experiment, return or dispose of the metal strips as directed by your instructor.

Table 26.2 Solution Combinations

Test Tube	Halogen Solution	Halide Solution
1	Cl_2	KBr
2	Cl_2	KI
3	Br_2	NaCl
4	Br_2	KI
5	I_2	NaCl
6	I_2	KBr

2. Halogens

A. Select a small test tube and pour 10 drops of KBr into it. Then add about half a dropperful of trichloroethane (TCE). Shake the test tube to make sure the TCE, which is heavier than water, settles to the bottom.

B. Now add 3–4 drops of chlorine water. Shake the test tube well to make sure the free halogen is transferred to the bottom layer. Allow the two phases to separate and note the color of the *bottom layer*. By the color of this layer—and this layer only!—determine which elemental halogen (Cl_2, Br_2, or I_2) is present. Knowing this, you can then use your reasoning (see examples above) to determine if a reaction has taken place.

C. Repeat the procedure, using the same order: (1) halide solution first, (2) TCE next, and (3) halogen last, adding KI and Cl_2 (test tube 2 in Table 26.2).

D. Continue mixing the remaining pairs (test tubes 3–6), one at a time, and note the color of the *bottom layer*. Record your observations on your work page.

RESULTS

Record the answers to the following questions in the space provided on the work page:

1. Which combination(s) yielded a redox reaction?
2. For each reaction, write the half-reaction equations for both oxidation and reduction.
3. If necessary, multiply either or both equations to equalize the electrons gained and lost in the half-reaction equations.
4. Add the half-reaction equations to get the net ionic redox equation and record them on the work page. If a reaction did not occur, write the equation and slash the arrow to show no reaction.
   Example: $Zn^{2+} + Cu \rightarrow Zn + Cu^{2+}$

5. From Part 1, list the oxidizers (oxidizing agents) in a column according to decreasing strength. Judge oxidizing strength by considering which species each oxidizer was capable of oxidizing and which species it could not oxidize. The last species will be the oxidizer that was incapable of oxidizing anything.

   Example: If metal A reacted with B^{2+} and C^{2+}, and metal B reacted with C^{2+} but not with A^{2+}, and metal C did not react with either A^{2+} or B^{2+}, then the strongest oxidizer is C^{2+} and the weakest oxidizer is A^{2+}. Remember, the oxidizers in this step are the metal ions.

6. Prepare a similar list for Part 2.

Experiment 26

Advance Study Assignment

1. Define the following terms:

   a. Reduction

   b. Half-reaction

   c. Oxidizer

2. Combine the following half-reactions to produce a balanced net ionic redox equation:

   a.
   $$A \rightarrow A^+ + e^-$$
   $$Z + 2\ e^- \rightarrow Z^{2-}$$

   b.
   $$B \rightarrow B^{2+} + 2\ e^-$$
   $$Y + 3\ e^- \rightarrow Y^{3-}$$

   c. Which specie in reaction a) is the reducing agent? Why?

Experiment 26

Work Page

Part 1—Metal–Metal Ion Reactions

Metal	Ion in Solution	Reaction Yes	Reaction No	Half-Reactions	Net Ionic Equations
Cu	Zn^{2+}				
	Pb^{2+}				
	Ag^+				
Zn	Cu^{2+}				
	Pb^{2+}				
	Ag^+				
Pb	Cu^{2+}				
	Zn^{2+}				
	Ag^+				
Ag	Cu^{2+}				
	Zn^{2+}				
	Pb^{2+}				

List of Oxidizers in Order of Decreasing Strength

_____

_____

_____

_____

Part 2—Halogens

Test Tube	Halide Solution	Halide Ion	Halogen	Color (after mixing)	Reaction	
					Yes	No
1	KBr	Br^-	Cl_2			
2	KI	I^-	Cl_2			
3	NaCl	Cl^-	Br_2			
4	KI	I^-	Br_2			
5	NaCl	Cl^-	I_2			
6	KBr	Br^-	I_2			

Half-Reactions and Net Ionic Redox Equations

List of Oxidizers in Order of Decreasing Strength

_____

_____

_____

Experiment 26

Report Sheet

Part 1—Metal–Metal Ion Reactions

Metal	Ion in Solution	Reaction		Half-Reactions	Net Ionic Equations
		Yes	No		
Cu	Zn^{2+}				
	Pb^{2+}				
	Ag^+				
Zn	Cu^{2+}				
	Pb^{2+}				
	Ag^+				
Pb	Cu^{2+}				
	Zn^{2+}				
	Ag^+				
Ag	Cu^{2+}				
	Zn^{2+}				
	Pb^{2+}				

List of Oxidizers in Order of Decreasing Strength

_____

_____

_____

_____

Part 2—Halogens

Test Tube	Halide Solution	Halide Ion	Halogen	Color (after mixing)	Reaction	
					Yes	No
1	KBr	Br^-	Cl_2			
2	KI	I^-	Cl_2			
3	NaCl	Cl^-	Br_2			
4	KI	I^-	Br_2			
5	NaCl	Cl^-	I_2			
6	KBr	Br^-	I_2			

Half-Reactions and Net Ionic Redox Equations

List of Oxidizers in Order of Decreasing Strength

_____

_____

_____

Nomenclature of Simple Organic Compounds: A Study Assignment

Performance Goal

27–1 Within the limits discussed in this exercise, give the name derived from the provided organic structure, or draw the structure from the name.

INTRODUCTION

This study assignment presents a brief introduction to International Union of Pure and Applied Chemistry (IUPAC) nomenclature rules and some common names. The rules selected here are by no means a comprehensive treatment of IUPAC nomenclature, and several areas have been neglected for the sake of simplicity. The areas not covered include: multifunctional organic compounds, complex substituents, stereochemistry, aromatic heterocyclic common names, and complex functional groups. Omitting polyfunctional organic nomenclature also eliminates the need to assign priorities to functional groups. The lab is designed with allied health students in mind and is not directed at chemistry majors.

The student must be familiar with stick diagrams to complete the lab.

CHEMICAL OVERVIEW

The IUPAC committee devised a way to systematically name organic molecules. This lab will outline how to name monofunctional organic molecules. Because of their frequency of use, some common names have also been added.

The process of assigning a name to an organic molecule has been broken down into a series of rules as outlined below.

ALKANES

Straight chain alkanes are named with a prefix that refers to the number of carbons and then adding the -ane suffix (see Table 27.1).

Branched alkanes and all other organic molecules are named according to a series of rules.

Table 27.1 *Nomenclature of straight chain alkanes*

Carbon Chain Length	Prefix	Alkane Name
1	meth-	methane
2	eth-	ethane
3	prop-	propane
4	but-	butane
5	pent-	pentane
6	hex-	hexane
7	hept-	heptane
8	oct-	octane
9	non-	nonane
10	dec-	decane

Rule 1. Find the longest contiguous chain in the molecule and name it.

A. If there are two chains of identical length, choose the one that contains the largest number of substituents.

> **Example 1**
>
>
>
> base is heptane   base is hexane
> 3-methyl-4-propylheptane   2,4,4-trimethylhexane

B. If a principal functional group is present, select the longest chain that includes that functional group.

Rule 2. Name all groups attached to the longest chain as substituents.

A. Alkyl substituents are named by dropping the -ane from the name and adding the suffix -yl. For example, methane becomes methyl, and ethane becomes ethyl (see Figure 27.1).

B. If there are identical substituents, number them and add the prefix di, tri, tetra, etc., to indicate the number of times that they appear.

C. Halogens are named as substituents such as fluoro-, chloro-, bromo-, and iodo-.

isopropyl   isobutyl   *tert*-butyl   *sec*-butyl

Figure 27.1
Common names of alkyl substituents

Example 2

2,5-dimethylnonane   3,4-diethyl-5-methylheptane

D. Some substituents are given common names as outlined in the table below.

Example 3

4-isopropyloctane   5-*tert*-butyl-3-methyloctane

Rule 3. Number the carbons of the longest chain, beginning with the end that is closest to a substituent.

A. In the case of two substituents being numbered the same, number according to alphabetical order.

B. If a principal functional group is present, assign the smallest number possible to that functional group.

C. A comma is used to separate numbers of identical substituents.

Example 4

2,4-dimethylpentane   1,2,4-triethylcyclohexane

Rule 4. Write the name of the principal chain. Then arrange all of the substituents in alphabetical order (each preceded by the carbon number to which it is attached and a hyphen) and then add the name of the stem. The prefixes referring to the number of substituents such as di-, tri-, tetra-, are

ignored in alphabetizing. The prefix iso- is not ignored but tert- and sec-
are.

Example 5

6-ethyl-3,4-dimethylnonane

5-ethyl-4-isopropyl-2-methylheptane

CYCLIC STRUCTURES

Rule 5. Cyclic structures are designated by the prefix cyclo-.

A. Numbering cyclic structures depends on the number of substituents on
the ring. If the structure contains two substituents, they are numbered
alphabetically. Three or more substituents are numbered by assigning
the order that gives the lowest possible number to all of the
substituents.

Example 6

1-ethyl-4-methylcyclohexane

1,4-diethyl-2-methylcyclopentane

B. If a principal functional group is present, it gets priority and is numbered first.

Example 7

1,6-dimethylcyclohex-1-ene

3-methylcyclopentanol

AROMATIC COMPOUNDS

Rule 6. Aromatic rings are named and numbered in the same fashion as cyclic structures.

A. However, common names are often used to denote benzene derivatives as shown in Figure 27.2.

| benzene | toluene | aniline | benzoic acid | phenol | benzyl aldehyde |

Figure 27.2
Common names of benzene derivatives

B. In addition, there is a method to denote the substitution pattern of disubstituted benzene rings by using the prefixes ortho-, meta-, and para-, which are abbreviated *o*-, *m*-, and *p*-. The substitution patterns are outlined in Figure 27.3.

Figure 27.3
Substitution patterns and nomenclature for disubstituted benzene

Example 8

1-ethyl-3-methylbenzene
or *m*-ethyltoluene

2-ethylaniline
or *o*-ethylaniline

2,5-dimethylphenol

FUNCTIONAL GROUPS

Rule 7. Suffixes identify the principal functional group or class to which the molecule belongs (see Table 27.1). Some functional groups such as amines, amides, ethers, and esters also utilize prefixes to denote the entire

Table 27.2 *Nomenclature rules for functional groups*

Functional Group	Suffix and Prefix	Functional Group	Suffix and Prefix
1. Alkane $R-\overset{\displaystyle R}{\underset{\displaystyle R}{C}}-\overset{\displaystyle R}{\underset{\displaystyle R}{C}}-R$	-ane	8. Ketone $\overset{O}{\overset{\|}{R-C-R}}$	-one
2. Alkene $\overset{R}{\underset{R}{}}C=C\overset{R}{\underset{R}{}}$	-ene	9. Carboxylic acid $\overset{O}{\overset{\|}{R-C-OH}}$	-oic acid
3. Alkyne $R-C\equiv C-R$	-yne	10. Ester $\overset{O}{\overset{\|}{R-C-OR'}}$	name R', main chain -oate
4. Halide $R-X$ $X = F,Cl,Br,I$	-o	11. Amine $R-N\overset{R'}{\underset{R'}{}}$	N-R' group, main chain, amine
5. Alcohol $R-OH$	-ol	12. Amide $\overset{O}{\overset{\|}{R-C-N}}\overset{R'}{\underset{R'}{}}$	N-R' group, main chain, -amide
6. Aldehyde $\overset{O}{\overset{\|}{R-C-H}}$	-al	13. Thiol $R-S-H$	main chain -thiol
7. Ether $R-O-R$	shorter chain –oxy, main chain		

functional group. All of the suffixes and prefixes for a select list of principal functional groups are listed in Table 27.2.

A. When numbering and naming a compound, the principal functional group takes precedence. Thus, the principal functional group is always included in the base chain and is numbered so that the principal functional group is assigned the lowest number possible.

B. Some functional groups are numbered in front of the suffix instead of the root word. This occurs when the functional group can appear at variable positions. Terminal functional groups, such as aldehydes, carboxylic acids, esters, and amides, are numbered 1 as long as they are the principal functional group. The number 1 is usually omitted for these groups.

C. When adding a suffix to the root, if two vowels appear together, the first vowel is dropped.

Example 9

5-methylhexan-3-one   4,4-dimethylpent-2-ene   3,3-dimethylhexanoic acid
(carboxylic acid is terminal
always number 1)

Example 10

N-ethyl-*N*-methyloctan-4-amine

2-methylcyclohexanol

3-ethylphenol
m-ethylphenol

2-ethylhexanal

5-methylhexan-2-one

4-ethyl-1,
2-dimethylcyclohexane

N-ethyl-*N*-methylpentanamide

1-ethoxypropane

5-chloro-4-methylhexanoic acid

Experiment 27

Advance Study Assignment

1. Write the names that correspond to the following structures.

Structure   Name

a.

b.

c.

2. Write the structures that correspond to the following names.

Name   Structure

a. 2-ethoxybutane

b. 3-methylcyclohexanone

c. 4-methylhexanal

Experiment 27

Work Page

General Instructions: For each substance whose name is given, write the structure; if the structure is given, write the name.

Alkanes

	2,2-dimethylpentane
	6-ethyl-4-isopropyl-3-methyloctane
	3-ethyl-3,4-dimethylhexane
	2,3,5-trimethylhexane
	4-*tert*-butyl-3,5-diethylheptane

Cyclic Structures

	1,1-dimethylcycloheptane
	1,2,3-trimethylcyclopropane
	1-ethyl-3-isopropylcyclopentane
	4-fluorocyclohex-1-ene
	1-ethyl-3-methylcyclobutane

Aromatic Compounds

	o-ethyltoluene
	3-chlorophenol

O OH ― C₆H₄ ― C(CH₃)₃ structure	1-ethyl-3,5-dimethylbenzene
CH₃ / Cl / CH₃ substituted benzene structure	*p*-isopropylbenzylaldehyde
dimethyl ethyl benzene structure	5-ethyl-2,3-dimethylphenol

Functional Groups

ketone structure	4-methylpent-2-yne
ether structure	methyl 2-methylbutanoate
ester structure	2-propylcyclopentanone
carboxylic acid structure	*N*-ethyl-*N*-methylethanamine

	N-methylpentanamide
	3-fluoro-4-methylpentanoic acid
	3,4-dimethylpentanal
	1-methoxypropane
	2,4-dimethylpentan-3-ol
	2-chloro-3,4,5,5-tetramethylhex-2-ene

Experiment 27

Report Sheet

General Instructions: For each substance whose name is given, write the structure; if the structure is given, write the name.

Alkanes

	2,2-dimethylpentane
	6-ethyl-4-isopropyl-3-methyloctane
	3-ethyl-3,4-dimethylhexane
	2,3,5-trimethylhexane
	4-*tert*-butyl-3,5-diethylheptane

Cyclic Structures

	1,1-dimethylcycloheptane
	1,2,3-trimethylcyclopropane
	1-ethyl-3-isopropylcyclopentane
	4-fluorocyclohex-1-ene
Cl	1-ethyl-3-methylcyclobutane

Aromatic Compounds

	o-ethyltoluene
NH$_2$	3-chlorophenol

	1-ethyl-3,5-dimethylbenzene
	p-isopropylbenzylaldehyde
	5-ethyl-2,3-dimethylphenol

Functional Groups

	4-methylpent-2-yne
	methyl 2-methylbutanoate
	2-propylcyclopentanone
	N-ethyl-*N*-methylethanamine

	N-methylpentanamide
	3-fluoro-4-methylpentanoic acid
	3,4-dimethylpentanal
	1-methoxypropane
	2,4-dimethylpentan-3-ol
	2-chloro-3,4,5,5-tetramethylhex-2-ene

Hydrocarbons and Alcohols

Performance Goals

28–1 Perform various tests on an alkane to determine whether it is saturated or unsaturated.

28–2 Perform tests on an alcohol to determine if it is a primary, secondary, or tertiary alcohol.

CHEMICAL OVERVIEW

Organic chemistry is the study of the compounds of carbon other than the carbonates, cyanides, carbon monoxide, and carbon dioxide. All organic molecules contain carbon and hydrogen; many of them also contain oxygen. Organic compounds are divided into subgroups based on common "functional groups." These subgroups have similar chemical properties and can be identified by laboratory tests.

Hydrocarbons

The simplest types of organic compounds are the *hydrocarbons*, which are composed of carbon and hydrogen only. In these compounds, carbon atoms may join in an open chain or form ring structures:

Propane (an alkane)   *Cyclohexene (an alkene)*

Saturated hydrocarbons are those in which each carbon is bonded to four other atoms, either hydrogens or other carbon atoms. Hydrocarbons in this class are named **alkanes** or **paraffins.** The first few members of alkanes, such as methane and propane, are gases; the higher-molar-mass alkanes, such as those found in paraffin wax, are solids.

Unsaturated hydrocarbons contain one or more double or triple bonds. Those with double bonds are **alkenes**, and those with triple bonds are

alkynes. Addition to a double or triple bond characterizes these compounds:

$$
\begin{array}{c}
\underset{|}{\overset{|}{C}}=\underset{|}{\overset{|}{C}} \ + \ Br_2 \ \longrightarrow \ \underset{|}{\overset{|}{C}}-\underset{|}{\overset{|}{C}} \\
 Br \quad Br
\end{array}
\tag{28.1}
$$

In general, alkenes and alkynes are much more reactive than alkanes are.

The reaction in Equation 28.1 is used to detect unsaturation. Bromine imparts a brownish color to a liquid. The disappearance of that color is a positive test for a multiple bond.

Unsaturated hydrocarbons can be oxidized by strong oxidizing agents, such as potassium permanganate, $KMnO_4$. Evidence of the reaction is the rapid disappearance of the purple color of the permanganate ion. Alcohols also react with $KMnO_4$.

Alcohols

Alcohols contain carbon, hydrogen, and oxygen, with the common functional group, —OH. This is known as the **hydroxyl group.** Depending on where the hydroxyl group is attached in a molecule, the alcohol is classified as primary (1°), secondary (2°), or tertiary (3°). In a primary alcohol the —OH group is connected to a carbon that is joined to only one other carbon atom. In a secondary alcohol the carbon that has the —OH group is joined to two other carbons, and in a tertiary alcohol the carbon is bonded to three other carbons.

$$
R_1-OH \qquad\qquad R_1-\underset{\underset{OH}{|}}{\overset{}{C}}-R_2 \qquad\qquad R_1-\underset{\underset{OH}{|}}{\overset{\overset{R_2}{|}}{C}}-R_3
$$

Primary $\qquad\qquad$ *Secondary* $\qquad\qquad$ *Tertiary*

In these diagrams, R_1, R_2, and R_3 represent alkyl groups. They may or may not be the same.

Alcohols are easily oxidized to either aldehydes or ketones:

$$
R-\underset{\underset{OH}{|}}{\overset{\overset{H}{|}}{C}}-H \ \overset{\text{ox.}}{\longrightarrow} \ R-\underset{\underset{O}{\|}}{\overset{}{C}}-H \qquad \text{(an aldehyde)}
\tag{28.2}
$$

$$
R_1-\underset{\underset{OH}{|}}{\overset{\overset{H}{|}}{C}}-R_2 \ \overset{\text{ox.}}{\longrightarrow} \ R_1-\underset{\underset{O}{\|}}{\overset{}{C}}-R_2 \qquad \text{(a ketone)}
\tag{28.3}
$$

Tertiary alcohols resist oxidation, even by strong oxidizing agents.

By oxidation and subsequent testing of the oxidation product, an alcohol can be classified as primary, secondary, or tertiary. The Lucas test will be used in this experiment to classify alcohols. Tertiary alcohols are converted to alkyl halides very rapidly, secondary alcohols take a few minutes, while primary alcohols react very slowly, with no observable change in 10 minutes.

$$R_1 - \underset{\underset{OH}{|}}{\overset{\overset{R_2}{|}}{C}} - R_3 \ + \ HCl \ \xrightarrow{\ ZnCl_2\ } \ R_1 - \underset{\underset{Cl}{|}}{\overset{\overset{R_2}{|}}{C}} - R_3 \ + \ H_2O \qquad (28.4)$$

(Insoluble in water)

Liquids of similar polarity tend to be miscible. Alcohols have —OH groups, which impart polar characteristics and the ability to hydrogen bond with water. As a result, most lower molar mass alcohols are water soluble.

Alkanes and alkenes are nonpolar so they are soluble in nonpolar or weakly polar organic solvents. Water is a polar solvent and is not miscible with any hydrocarbon.

SAFETY PRECAUTIONS AND DISPOSAL METHODS

The organic chemicals you will use in this experiment are highly volatile and flammable. Do not breathe the vapors and be sure not to use them near flames. Bromine, Br_2, causes severe burns on contact with the skin. Handle this reagent very carefully and only in the fume hood. Avoid breathing the vapors, which are also very harmful. Be sure to wear eye protection while performing this experiment, and wash your hands when you are finished.

Dispose of the organic liquids in a special container provided or as directed by your instructor. *Do not pour organic liquids down the sinks in the open laboratory!*

PROCEDURE

1. Solubility of Alcohols and Hydrocarbons

Add 5 mL of water to five dry, labeled test tubes. Now pour 1 mL of each of the following liquids into separate test tubes and shake. Observe if the substance dissolves and record your observations on the work page.

Test tube No. 1: ethanol (CH_3CH_2OH)

Test tube No. 2: 2-propanol ($CH_3CHOHCH_3$)

Test tube No. 3: *t*-butanol ((CH_3)$_3COH$)

Test tube No. 4: cyclohexene (C_6H_{10})

Test tube No. 5: hexane (C_6H_{14})

2. Reactions of Alcohols and Hydrocarbons

A. Reaction with Br₂. Add 1 mL hexane (an alkane) and 1 mL cyclohexene (an alkene) to two separate, dry, labeled test tubes. Add 2 to 3 drops of bromine solution to each and shake. The disappearance of a brownish color is indicative of an alkene. Record your results on the work page.

B. Reaction with KMnO₄. Add 1 mL hexane and 1 mL cyclohexene to two separate, dry, labeled test tubes. Add 2 to 3 drops of KMnO₄ solution to each and shake. The disappearance of the purple color of KMnO₄ with the appearance of a brown precipitate (MnO₂) is a positive test for unsaturated hydrocarbons. Record your results on the work page.

C. Lucas test. Add 1 mL of ethanol, 1 mL 2-propanol, and 1 mL *t*-butyl alcohol to three separate, dry, labeled test tubes. Pour 3 mL of Lucas reagent into each, insert stopper, and shake to mix. Allow the solutions to stand at room temperature for up to 15 minutes. The appearance of cloudiness indicates a reaction. If no change is visible, immerse the test tube(s) into a water bath maintained at approximately 60°C, and observe whether any change occurs. Discontinue heating after 15 minutes. Record your observations on the work page.

3. Identification of an Unknown

A. Solubility. Obtain an unknown from your instructor and record the unknown number. Add 5 mL of water to a dry test tube. Pour 1 mL of the unknown into the test tube and shake. Observe if the substance dissolves and record your observations. If the unknown is insoluble, go to part B. If it is soluble, go to part C.

B. Reaction with Br₂ and KMnO₄. Add 1 mL of the unknown to two separate, dry, labeled test tubes. Add 2 to 3 drops of bromine solution to one test tube, and add 2 to 3 drops of KMnO₄ solution to the other test tube. Shake both test tubes and observe any color change.

C. Lucas test. Pour 1 mL of the unknown into a dry test tube and perform the Lucas test.

D. Identify the unknown as ethanol, 2-propanol, *t*-butanol, cyclohexene, or hexane.

Experiment 28

Advance Study Assignment

1. Predict the solubility of the following compounds in the specified solvent by writing an M for miscible or NM for nonmiscible in the following table.

	$CH_3(CH_2)_4CH_3$ (**Hexane**) Solvent	H_2O (**Water**) Solvent
CH_3CH_2OH		
$CH_3CH_2CH_2CH_3$		
$CH_3CHOHCH_3$		

2. **A.** Write the equation for the reaction of cyclohexene with bromine.

B. Predict the color or appearance of a solution if two drops of bromine are added to cyclohexene.

3. Complete the table below using the information given in the Chemical Overview sections. In each blank space in the table write S (slow), F (fast), or NR (no reaction).

Lucas Test

Type of Alcohol	Room Temp.	60°C
1°		
2°		
3°		

Experiment 28

Work Page

Part 1—Solubility of Alcohols and Hydrocarbons

Test Tube	Substance	Solubility
1		
2		
3		
4		
5		

Part 2—Reactions of Alcohols and Hydrocarbons

Step A—Reaction of Br₂ with Alkenes

Substance	Yes	No
Hexane		
Cyclohexene		

Step B—Reaction of KMnO₄ with Alkenes

Substance	Yes	No
Hexane		
Cyclohexene		

If the structure of acetylene is $HC \equiv CH$, would you predict that it gives a positive _____ or negative _____ test with $KMnO_4$?

Step C—Lucas Test

Test Tube	Substance	Lucas Test		Time (min)	
		Yes	No	Room Temp.	60°C
1					
2					
3					

Part 3—Identification of an Unknown

Unknown number _____

Step A—Solubility

Soluble in H_2O?   Yes _____ No _____

If not soluble go to step B, if soluble go to step C.

Step B—Reactions of Alkenes

Reaction with Br_2?   Yes _____ No _____

Reaction with $KMnO_4$?   Yes _____ No _____

Step C—Lucas Test

Lucas test was positive at room temperature _____, or 60°C _____.

Identity of unknown _____

Experiment 28

Report Sheet

Part 1—Solubility of Alcohols and Hydrocarbons

Test Tube	Substance	Solubility
1		
2		
3		
4		
5		

Part 2—Reactions of Alcohols and Hydrocarbons

Step A—Reaction of Br₂ with Alkenes

Substance	Yes	No
Hexane		
Cyclohexene		

Step B—Reaction of KMnO₄ with Alkenes

Substance	Yes	No
Hexane		
Cyclohexene		

If the structure of acetylene is $HC \equiv CH$, would you predict that it gives a positive _____ or negative _____ test with $KMnO_4$?

Step C—Lucas Test

Test Tube	Substance	Lucas Test		Time (min)	
		Yes	No	Room Temp.	60°C
1					
2					
3					

Part 3—Identification of an Unknown

Unknown number _____

Step A—Solubility

Soluble in H_2O?   Yes _____ No _____

If not soluble go to step B, if soluble go to step C.

Step B—Reactions of Alkenes

Reaction with Br_2?   Yes _____ No _____

Reaction with $KMnO_4$?   Yes _____ No _____

Step C—Lucas Test

Lucas test was positive at room temperature _____, or 60°C _____.

Identity of unknown _____

Aldehydes, Ketones, and Carboxylic Acids

Performance Goal

29–1 Identify an unknown as an aldehyde, ketone, or carboxylic acid.

CHEMICAL OVERVIEW

Organic compounds are best studied according to the functional groups they contain. Aldehydes and ketones both contain the carbonyl group,

$$\diagdown \!\!\! \diagup \!\! C = O$$

In aldehydes, one of the bonds is always connected to a hydrogen atom. Thus all aldehydes have a

$$\underset{H}{\overset{\diagdown}{\diagup}} C = O$$

group. All carboxylic acids contain the carboxyl group,

$$-C \!\!\! \underset{OH}{\overset{O}{\diagup \!\! \diagdown}}$$

Aldehydes and Ketones

When a primary (1°) alcohol is oxidized, the product is an aldehyde, while a secondary (2°) alcohol yields a ketone. The —CHO group of an aldehyde is easily oxidized to a carboxylic acid:

$$R - CHO \xrightarrow{\text{ox.}} R - COOH \qquad (29.1)$$

Aldehyde   *Carboxylic acid*

379

Ketones, on the other hand, resist oxidation. This difference is the basis of several common laboratory tests that distinguish aldehydes from ketones. Two such tests use metal ions as the oxidizing agent:

$$R\text{—}CHO \xrightarrow{Ag^+} Ag\,(s) + R\text{—}COOH \qquad \text{Tollen's test} \qquad (29.2)$$

$$R\text{—}CHO \xrightarrow{Cu^{2+}} Cu_2O\,(s) + R\text{—}COOH \qquad \text{Benedict's test or Fehling's test} \qquad (29.3)$$

A very common way to identify a ketone with an $R\text{—}\overset{\displaystyle O}{\overset{\|}{C}}\text{—}CH_3$ group is the iodoform test. Iodoform, CHI_3, is a yellow solid with a melting point of 119°C. The ketone reacts with I_2 in an alkaline solution to give CHI_3:

$$R\text{—}\overset{\displaystyle O}{\overset{\|}{C}}\text{—}CH_3 \xrightarrow{I_2,\,NaOH} R\text{—}COO^- + CHI_3\,(s) \qquad (29.4)$$

Iodoform has a characteristic medicine-like smell, which helps to establish its presence.

The lower-molar-mass aldehydes and ketones are water soluble, but as the carbon chain lengthens, they become insoluble.

Carboxylic Acids

Most organic acids are weak acids because they ionize only slightly in aqueous solutions. For example, 1 molar acetic acid is only about 1 percent ionized; 99 percent of the compound remains in molecular form. Even though the hydrogen ion concentration is lower than it would be for the same concentration of a strong acid, such as HCl, HNO_3, or H_2SO_4, the solutions of organic acids *are* acidic. They have low pH values, whereas aldehydes and ketones are neutral.

Organic acids, except for those with low molar mass, are not soluble in water. Upon reaction with a strong base (NaOH), they form salts that *are* water soluble.

$$R\text{—}\overset{\displaystyle O}{\underset{OH}{C}}\!\!\diagup + OH^- \longrightarrow R\text{—}\overset{\displaystyle O}{\underset{O^-}{C}}\!\!\diagup + H_2O \qquad (29.5)$$

A typical acid-base reaction occurs between carboxylic acids and aqueous $NaHCO_3$. Gaseous CO_2 is produced according to

$$R\text{—}\overset{\displaystyle O}{\underset{OH}{C}}\!\!\diagup + HCO_3^- \longrightarrow R\text{—}\overset{\displaystyle O}{\underset{O^-}{C}}\!\!\diagup + H_2O + CO_2\,(g) \qquad (29.6)$$

Neither aldehydes nor ketones react with aqueous $NaHCO_3$.

When a carboxylic acid reacts with an alcohol, the product is called an ester (see Experiment 31). Many esters have distinct odors (oil of wintergreen, banana oil, apricot oil, etc.) and are widely used as scents or flavorings.

SAFETY PRECAUTIONS AND DISPOSAL METHODS

Several organic compounds used in this experiment are volatile and have very strong odors. Handle these chemicals in the hood and be careful not to get them on your skin. Acetone is flammable; do not use it near flames. Avoid breathing the vapors of chemicals; some are potentially harmful. Be sure to wear eye protection while performing this experiment.

Dispose of the organic liquids in a special container provided or as directed by your instructor. *Do not pour organic liquids down the sinks!*

PROCEDURE

1. Aldehydes and Ketones

A. Benedict's test. Label three clean, dry test tubes and pour 1 mL of formaldehyde solution, propanal, and acetone into separate test tubes. Prepare a fourth tube with deionized (distilled) water (to be used as a blank). Add to each test tube 5 mL of Benedict's reagent. Place the test tubes in a boiling-water bath for 2 to 3 minutes. The formation of a reddish-brown precipitate (Cu_2O) indicates the presence of an aldehyde. To be considered a positive test, a precipitate *must* be present. Enter your observations on the work page.

B. Iodoform test. Label three large test tubes and add 2 mL of water to each. Now add 5 drops of formaldehyde solution, propanal, and acetone to separate test tubes. Pour 1 mL of 10% NaOH solution into each and add dropwise, with shaking, enough iodine test solution to give a definite, dark iodine color. If at this point less than 2 mL (about 40 drops) of iodine solution was used, heat the test tube in a water bath at about 60°C. If the solution is decolorized on heating, add more iodine test solution until the dark color persists for 2 minutes at 60°C. Now add a few drops of 10% NaOH while shaking to expel any unreacted iodine. The solutions at this point will remain yellow.

Fill each test tube with deionized water and allow it to stand for 15 minutes. (Proceed with the next part of the experiment while waiting.) If a methyl ketone was present, a yellow precipitate (CHI_3) will form. Record your observations on the work page.

2. Carboxylic Acids

A. pH test. Pour 5 mL of deionized water into four labeled test tubes. Add 1 mL of acetic acid to the first test tube, 1 mL of propionic acid (also called propanoic acid) to the second, and 0.1 g of salicylic acid to the third. The fourth test tube will serve as a blank. If the acid does not dissolve, use ethanol as a solvent instead of water and then add water until crystals just begin to form. Test the pH of each solution with a pH paper and record your observations on the work page.

B. Sodium hydrogen carbonate test. To three labeled test tubes add 3 mL of 10% sodium hydrogen carbonate ($NaHCO_3$) solution. To the first one add 10 drops of acetic acid; to the second, 10 drops of propionic acid; and to the third, 0.1 g of salicylic acid. A distinct fizzing sound or visual observation of the formation of CO_2 bubbles constitutes a positive test.

3. Identification of an Unknown

Obtain an unknown from your instructor. If it is a solid, use 0.1-g portions; if it is a liquid, use the same amount you used for the knowns. Repeat tests A and B for aldehydes and ketones, followed by the tests for carboxylic acids. Based on your observations, classify your unknown as an aldehyde, ketone, or acid. Enter your observations on the work page.

Experiment 29

Advance Study Assignment

1. Answer the following questions as true (T) or false (F):

   a. Organic acids are weak acids.   _____

   b. The pH of 0.10 M HCl is higher than the pH of 0.10 M acetic acid.   _____

   c. Sodium salts of organic acids are soluble in water.   _____

   d. Both aldehydes and ketones contain the carbonyl group.   _____

2. Describe the difference in behavior between ketones and carboxylic acids when reacting with sodium hydrogen carbonate.

3. If an unknown compound in this experiment gave a negative Benedict's test, a positive iodoform test, and a negative sodium hydrogen carbonate test, it would be a(n) (aldehyde, ketone, or carboxylic acid):

Experiment 29

Work Page

PART 1—Aldehydes and Ketones

Step A—Benedict's Test

Test Tube	Substance	Yes	No
1	Formaldehyde		
2	Propanal		
3	Acetone		
4	H_2O		

Step B—Iodoform Test

Test Tube	Substance	Room Temp.		60°C	
		Yes	No	Yes	No
1	Formaldehyde				
2	Propanal				
3	Acetone				

PART 2—Carboxylic Acids

Step A—pH Test

Compare results obtained on the three knowns to the pH of your deionized (distilled) water.

Test Tube	Substance	Approx. pH
1	Acetic acid	
2	Propionic acid	
3	Salicylic acid	

Are the solutions acidic _____, neutral _____, or basic _____?

Step B—Sodium Hydrogen Carbonate Test

Test Tube	Substance	Yes	No
1	Acetic acid		
2	Propionic acid		
3	Salicylic acid		

PART 3—Unknown

Unknown number _____

Benedict's test: Yes _____   No _____

Iodoform test: Yes _____   No _____

pH test: Acidic _____   Neutral _____   Basic _____

Sodium hydrogen carbonate test: Yes _____   No _____

Compound is an aldehyde _____, a ketone _____, an acid _____.

Experiment 29

Report Sheet

PART 1—Aldehydes and Ketones

Step A—Benedict's Test

Test Tube	Substance	Yes	No
1	Formaldehyde		
2	Propanal		
3	Acetone		
4	H_2O		

Step B—Iodoform Test

Test Tube	Substance	Room Temp.		60° C	
		Yes	No	Yes	No
1	Formaldehyde				
2	Propanal				
3	Acetone				

PART 2—Carboxylic Acids

Step A—pH Test

Compare results obtained on the three knowns to the pH of your deionized (distilled) water.

Test Tube	Substance	Approx. pH
1	Acetic acid	
2	Propionic acid	
3	Salicylic acid	

Are the solutions acidic _____, neutral _____, or basic _____?

Step B—Sodium Hydrogen Carbonate Test

Test Tube	Substance	Yes	No
1	Acetic acid		
2	Propionic acid		
3	Salicylic acid		

PART 3—Unknown

Unknown number _____

Benedict's test: Yes _____   No _____

Iodoform test: Yes _____   No _____

pH test: Acidic _____   Neutral _____   Basic _____

Sodium hydrogen carbonate test: Yes _____   No _____

Compound is an aldehyde _____, a ketone _____, an acid _____.

Esters

Performance Goals

30–1 Perform reactions between a series of carboxylic acids and alcohols to form esters.

30–2 Identify an unknown ester by its specific odor.

CHEMICAL OVERVIEW

Esters are functional groups of the configuration shown below where R and R′ represent alkyl groups (hydrocarbon). Many esters have pleasant aromas, which can be found in natural foodstuffs such as banana oil, pineapples, raspberries, oil of wintergreen, and oranges. Because of their pleasant odor, esters are often added into household products, cleaners, and foods to give them a specific scent. Helping their appeal as odor additives, esters are relatively volatile, meaning that they readily go into the gaseous state where they can be smelled.

$$\begin{matrix} & O & \\ & \| & \\ R-C & & O-R' \end{matrix}$$

One of the most common ways to synthesize esters is to cause a reaction between an alcohol and a carboxylic acid in the presence of an acid. The acid acts as a catalyst to increase the rate of reaction. This reaction produces an ester and a water molecule.

$$R-\overset{\overset{\displaystyle O}{\|}}{C}-OH \;+\; HO-R' \;\underset{\text{Heat}}{\overset{H^+}{\rightleftharpoons}}\; R-\overset{\overset{\displaystyle O}{\|}}{C}-O-R' \;+\; H_2O \quad (30.1)$$

A specific example would be the formation of methyl butyrate from the carboxylic acid (butyric acid) and the alcohol methyl alcohol. Butyric acid is associated with the smell of "rancid butter" or "sweat" and the resulting ester smells like "apple."

$$H_3CH_2CH_2C\overset{\overset{O}{\|}}{-}C-OH \; + \; HO-CH_3 \; \underset{}{\overset{H^+}{\rightleftharpoons}} \; H_3CH_2CH_2C\overset{\overset{O}{\|}}{-}C-CH_3 \; + \; H_2O \quad (30.2)$$

butyric acid   methyl alcohol   methyl butyrate "apple"

In this experiment, several esters will be formed from their corresponding alcohol and carboxylic acid. The resulting ester will be identified by its particular odor.

SAFETY PRECAUTIONS AND DISPOSAL METHODS

This experiment requires the use of a concentrated sulfuric acid. If this acid comes into contact with the skin, dab with a paper towel and then rinse the area with plenty of cold water. Be careful when using a water bath. Handle test tubes in the hot-water bath by using a test-tube holder. When smelling chemicals, make sure to waft the fumes toward your nose.

Dispose of the resulting esters in the containers provided.

PROCEDURE

1. Formation of Octyl Acetate

A. Add 15 drops of 1-octanol and 3 drops of 6 M sulfuric acid to a test tube containing 15 drops of concentrated acetic acid.

B. Mix the solution thoroughly.

C. Place the test tube containing the solution in a hot-water bath and heat.

D. Periodically take the test tube out of the water bath and note the aroma of the solution. The reaction should be complete after approximately 5 minutes. Be careful to waft the fumes toward your nose.

2. Formation of the Remaining Esters

A. Follow the table below by putting 15 drops of an alcohol and 15 drops of the paired carboxylic acid into a test tube. Add 3 drops of sulfuric acid and heat as outlined in Procedure 1. Note and record the odor of the resulting product.

B. Note that salicylic acid is a solid and not a liquid. To dissolve the acid, add a small scoop of salicylic acid to 15 drops of ethyl alcohol. Then add 3 drops of 6 M sulfuric acid, and then add the methanol.

3. Unknown Test

Obtain an unknown from your instructor and record the unknown number. Place 3 drops of 6 M sulfuric acid and approximately 5 mL of an unknown into the test tube, and heat in the water bath for at least 5 minutes. Determine the identity of your ester by noting and recognizing its particular odor.

Experiment 30

Advance Study Assignment

1. Write the reaction of pentanoic acid with 1-butanol in the presence of an acid.

2. What is the purpose of adding acid to the synthesis of esters?

3. What alcohol and carboxylic acid are used to form propyl hexanoate?

4. What molecule, besides an ester, is formed when an alcohol and a carboxylic acid react to form an ester?

Experiment 30

Work Page

Part 1—Formation of Esters

Carboxylic acid	Alcohol	Structure of resulting ester	Expected odor	Odor of product
O ‖ H₃C—C—OH Acetic acid	HO(CH₂)₇CH₃ Octanol		Orange peelings	
O ‖ C—OH (benzene ring with OH) Salicyclic acid	H₃C—OH Methanol		Wintergreen	
O ‖ H₃C—C—OH Acetic acid	HO(CH₂)₄CH₃ Pentanol		Banana oil	
O ‖ H—C—OH Formic acid	CH₃ | HOH₂C—C—CH₃ | H Isobutyl alcohol		Raspberry	
O ‖ H₃C—C—OH Acetic acid	HOH₂C—(benzene ring) Benzyl alcohol		Jasmine	

O $\parallel$ $H_3CH_2C-C-OH$ Propanoic acid	$HOCH_2CH_3$ Ethanol		Rum	
O $\parallel$ $H_3CH_2CH_2C-C-OH$ Butyric acid	H_3C-OH Methanol		Apple	
O $\parallel$ $H_3CH_2CH_2C-C-OH$ Butyric acid	$HOCH_2CH_3$ Ethanol		Pineapple	
O $\parallel$ $H_3CH_2CH_2C-C-OH$ Butyric acid	$HOH_2C-\langle\bigcirc\rangle$ Benzyl alcohol		Cherry	

Part 2—Identity of an Unknown Ester

Unknown number _____

Odor of unknown:

Draw the structure of the unknown.

Experiment 30

Report Sheet

Part 1—Formation of Esters

Carboxylic acid	Alcohol	Structure of resulting ester	Expected odor	Odor of product
$H_3C-\overset{\overset{\displaystyle O}{\|\|}}{C}-OH$ Acetic acid	$HO(CH_2)_7CH_3$ Octanol		Orange peelings	
Salicyclic acid	H_3C-OH Methanol		Wintergreen	
$H_3C-\overset{\overset{\displaystyle O}{\|\|}}{C}-OH$ Acetic acid	$HO(CH_2)_4CH_3$ Pentanol		Banana oil	
$H-\overset{\overset{\displaystyle O}{\|\|}}{C}-OH$ Formic acid	$HOH_2C-\overset{\overset{\displaystyle CH_3}{\|}}{\underset{\underset{\displaystyle H}{\|}}{C}}-CH_3$ Isobutyl alcohol		Raspberry	
$H_3C-\overset{\overset{\displaystyle O}{\|\|}}{C}-OH$ Acetic acid	HOH_2C- Benzyl alcohol		Jasmine	

H_3CH_2C—$\overset{\overset{\displaystyle O}{\|}}{C}$—OH Propanoic acid	$HOCH_2CH_3$ Ethanol		Rum	
$H_3CH_2CH_2C$—$\overset{\overset{\displaystyle O}{\|}}{C}$—OH Butyric acid	H_3C—OH Methanol		Apple	
$H_3CH_2CH_2C$—$\overset{\overset{\displaystyle O}{\|}}{C}$—OH Butyric acid	$HOCH_2CH_3$ Ethanol		Pineapple	
$H_3CH_2CH_2C$—$\overset{\overset{\displaystyle O}{\|}}{C}$—OH Butyric acid	HOH_2C—⬡ Benzyl alcohol		Cherry	

Part 2—Identity of an Unknown Ester

Unknown number _____

Odor of unknown:

Draw the structure of the unknown.

Preparation of Aspirin

Performance Goal

31–1 Beginning with salicylic acid and acetic anhydride, prepare a sample of aspirin.

CHEMICAL OVERVIEW

Chemically speaking, aspirin is an organic ester. An ester is a compound that is formed when an acid reacts with an alcohol (or a compound containing an —OH group):

$$\text{Acid} \qquad \text{Alcohol} \qquad \text{Ester} \qquad (31.1)$$

where R_1 and R_2 represent alkyl or aryl groups, such as CH_3—, C_2H_5—, or C_6H_5—.

High-molar-mass esters such as aspirin are generally insoluble in water and can be separated from a reaction mixture by crystallization. Aspirin can be prepared by the reaction of salicylic acid with acetic acid:

$$\text{Acetic acid} \qquad \text{Salicylic acid} \qquad \text{Aspirin} \qquad (31.2)$$

As the double arrow indicates, the reaction does not go to completion, but reaches equilibrium.

A better preparative method—the one you will use in this experiment—employs acetic anhydride instead of acetic acid. Acetic anhydride may be considered as the product of a reaction in which two acetic acid molecules combine, with the resulting elimination of a water molecule:

$$CH_3-C\!\!\diagup\!\!\overset{O}{\diagdown}_{OH} + \overset{O}{\diagup}\!\!\diagdown\!C-CH_3 \longrightarrow CH_3-C\!\!\diagup\!\!\overset{O\ O}{\diagdown}_{O}\diagdown C-CH_3 + H_2O \qquad (31.3)$$

The anhydride reacts with salicylic acid to yield the ester (aspirin):

$$CH_3-C\!\!\diagup\!\!\overset{O\ O}{\diagdown}_{O}\diagdown C-CH_3 + 2\ HO-\!\!\overset{O=C\diagup^{OH}}{\bigcirc} \longrightarrow 2\ CH_3-C\!\!\diagup\!\!\overset{O\ O=C\diagup^{OH}}{\diagdown}_{O}-\bigcirc + H_2O \qquad (31.4)$$

Acetic anhydride　　*Salicylic acid*　　*Aspirin*
(MM = 102)　　*(MM = 138)*　　*(MM = 180)*

Excess anhydride reacts with the water produced in the esterification, thereby shifting the equilibrium in the forward direction and giving a better yield of the desired product. A catalyst, normally sulfuric or phosphoric acid, is used to increase the rate of the reaction.

SAFETY PRECAUTIONS AND DISPOSAL METHODS

Both acetic anhydride and phosphoric acid are reactive chemicals that can produce a serious burn on contact with the skin. In case of contact with either, wash the skin thoroughly with soap and water. Avoid breathing acetic anhydride vapors. Wash any spillage from the desk top. The aspirin you will prepare in this experiment is relatively impure and should not be taken internally.

Dispose of any excess solid chemical in a special container. Do not pour acetic anhydride down the drain. Follow the directions given by your instructor.

PROCEDURE

1. Preparation of Aspirin

A. Preweigh a 50-mL Erlenmeyer flask on a centigram balance. Add 1.9 to 2.2 g of salicylic acid and weigh the flask again to the nearest 0.01 g.

B. Pour 5.0 to 5.5 mL of acetic anhydride into the flask in such a way as to wash down any crystals of salicylic acid that may have adhered to the walls.

C. Add 5 drops of concentrated phosphoric acid (85 percent) to serve as a catalyst.

Figure 31.1
Preparation of aspirin

D. Clamp the flask in a beaker of water supported on a wire gauze (Figure 31.1), or use a hot plate. Heat the water to about 75 °C, stirring the liquid in the flask occasionally with a stirring rod. Maintain this temperature for about 15 minutes, during which time the reaction should be complete.

E. *Cautiously* add 2 mL of water to the flask to decompose any excess acetic anhydride. Hot acetic acid vapor will evolve as a result of the decomposition.

F. When the liquid has stopped giving off vapors, remove the flask from the water bath and add 18 to 20 mL of water. Let the flask cool for a few minutes, during which time crystals of aspirin should begin to appear. Put the flask into an ice bath to hasten crystallization and increase the yield of the product. If crystals are slow to appear, it may be helpful to scratch the inside of the flask with a stirring rod.

G. Weigh a piece of filter paper on a centigram balance before inserting it into the funnel. Collect the aspirin by filtering the cold liquid through a Büchner funnel, using suction, as in Figure 31.2. Disconnect the rubber hose from the filter flask, pour about 5 mL of ice-cold deionized water over the crystals, and suck down the wash water. Repeat the washing step with a second 5-mL rinse of ice-cold water. Draw air through the funnel for a few minutes to help dry the crystals, and then transfer the filter paper and crystals to a clean watch glass.

Figure 31.2
Vacuum filtration apparatus

H. To determine the yield of aspirin in your experiment, it is necessary that the product be dry. If you do not have time to complete the experiment, store the watch glass carefully in your locker. At the beginning of the next laboratory period, weigh the filter paper and aspirin to the nearest 0.01 g. Record your data in the space provided on the work page.

2. Purity of Aspirin (Optional)

Very pure aspirin melts at 135°C. By determining the melting point of your aspirin, you may estimate its purity, because the purer the aspirin, the closer its melting point will be to 135°C.

A. Assemble the apparatus shown in Figure 31.3, using a large oil-filled test tube as the heating bath.

B. Crush some of your aspirin crystals on a watch glass with a spatula. Form a mound from the powder and push the open end of a melting-point capillary into the mound. Hold the capillary vertically and allow it to drop against the table top, compacting the powder into a plug in the bottom of the tube. Repeating the process, build a plug about $\frac{3}{4}$ to 1 cm long.

C. Attach the filled capillary to a thermometer with a rubber band or slice of rubber tubing, and immerse it in the oil bath. Do not allow the open end of the capillary to come into contact with the oil. Heat the bath rapidly with a Bunsen burner to about 100°C. As the melting point is approached, the crystals will begin to soften. Report the melting point as the temperature at which the last crystals disappear (the tube looks transparent).

Thermometer

Split
stopper

Large
test tube

Rubber band

Oil bath

Solid
crystal

Figure 31.3
*Apparatus for melting-point
determination*

Bunsen burner

CALCULATIONS

The actual mass of aspirin is obtained by taking the mass of filter paper +
aspirin and subtracting the mass of the filter paper. Based on the actual
mass of salicylic acid used, calculate the theoretical yield of aspirin in
grams, using Equation 31.4. Then determine the percentage yield,

$$\text{Percentage yield} = \frac{\text{actual yield}}{\text{theoretical yield}} \times 100$$

where *actual yield* means the number of grams of product actually
obtained. Record your results on the work page.

Experiment 31

Advance Study Assignment

1. How would you test the purity of aspirin prepared in this experiment?

2. Calculate the theoretical yield of aspirin if you started with 1.75 g of salicylic acid.

3. Write the chemical equation for the chemical reaction used in this experiment to form aspirin.

Experiment 31

Work Page

Part 1 —Preparation of Aspirin

1A. Mass of 50-mL Erlenmeyer flask (g)   _____

1B. Mass of flask and salicylic acid (g)   _____

1C. Mass of salicylic acid (g)   _____

1D. Mass of filter paper (g)   _____

1E. Mass of filter paper and aspirin (g)   _____

1F. Mass of aspirin (actual yield) (g)   _____

Theoretical yield of aspirin (show calculations below):

_____ g

Percentage yield (show calculations below):

_____ %

Part 2—Purity of Aspirin (Optional)

Melting point of aspirin   _____ °C

Experiment 31

Report Sheet

Part 1 —Preparation of Aspirin

1A. Mass of 50-mL Erlenmeyer flask (g)   _____

1B. Mass of flask and salicylic acid (g)   _____

1C. Mass of salicylic acid (g)   _____

1D. Mass of filter paper (g)   _____

1E. Mass of filter paper and aspirin (g)   _____

1F. Mass of aspirin (actual yield) (g)   _____

Theoretical yield of aspirin (show calculations below):

_____ g

Percentage yield (show calculations below):

_____ %

Part 2—Purity of Aspirin (Optional)

Melting point of aspirin   _____ °C

Preparation and Properties of a Soap

Performance Goals

32–1 Starting with a vegetable oil, prepare a soap in the laboratory.

32–2 Examine the chemical properties of the soap you prepared.

CHEMICAL OVERVIEW

An ester is the product of the reaction between an alcohol and a carboxylic acid. The typical equation for the formation of an ester is

$$R_1 - C \underset{OH}{\overset{O}{\diagup}} \ + \ HO - R_2 \ \longrightarrow \ R_1 - C \underset{O - R_2}{\overset{O}{\diagup}} \ + \ H_2O \tag{32.1}$$

Acid   *Alcohol*   *Ester*

where R_1 and R_2 are general symbols for *alkyl* groups containing only hydrogen and carbon. They may be the same group or they may be different. If the alcohol is glycerol, $C_3H_5(OH)_3$, and the acid is a long-chain fatty acid such as stearic acid, $C_{17}H_{35}COOH$, the ester is typical of those found in fats and oils. These esters can be reacted with strong bases to yield glycerol and the salt of the fatty acid. This process is known as *saponification,* and the sodium (or potassium) salts of the fatty acids are called *soaps.* The soap-making process may be written as

$$
\underbrace{
\begin{array}{l}
\text{H} \\
| \\
\text{H—C—O—C—R}_1 \\
\quad\quad\quad\;\; \| \\
\quad\quad\quad\;\; \text{O} \\
| \\
\text{H—C—O—C—R}_2 \\
\quad\quad\quad\;\; \| \\
\quad\quad\quad\;\; \text{O} \\
| \\
\text{H—C—O—C—R}_3 \\
\quad\quad\quad\;\; \| \\
\quad\quad\quad\;\; \text{O} \\
| \\
\text{H}
\end{array}
}_{\text{Ester}}
+ \underbrace{3\text{NaOH}}_{\text{Base}} \longrightarrow
\underbrace{
\begin{array}{l}
\text{H} \\
| \\
\text{H—C—O—H} \\
| \\
\text{H—C—O—H} \\
| \\
\text{H—C—O—H} \\
| \\
\text{H}
\end{array}
}_{\text{Glycerol}}
+ \underbrace{
\begin{array}{l}
\text{Na}^+{}^-\text{O—C—R}_1 \\
\|\\
\text{O}\\
\text{Na}^+{}^-\text{O—C—R}_2 \\
\|\\
\text{O}\\
\text{Na}^+{}^-\text{O—C—R}_3
\end{array}
}_{\text{Soap}}
\quad (32.2)
$$

As you can see from Equation 32.2, the anions of soaps contain both polar groups ($-\text{COO}^-$) and nonpolar groups (long chains of carbon atoms). Polar compounds (or groups) are attracted to water and are called **hydrophilic.** Nonpolar compounds (or groups) are water repelling or **hydrophobic,** and are soluble in or miscible with nonpolar compounds, such as fats, grease, oil, or other "dirt." This dual characteristic of soaps is the reason behind their cleaning action. The fat or oil is displaced from the fiber by the soap solution to form large globules that can be detached by jarring (rubbing) and then dispersed (emulsified) in the aqueous solution. **Emulsions** consist of fine droplets of one liquid dispersed in another liquid, in which the first is insoluble (like oil in water). The better the emulsion characteristics of a soap, the better it can clean.

Generally, soaps made from liquid fats (or oils) are more soluble than those made from solid fats. In the laboratory, you will prepare a soap by saponifying a vegetable oil with sodium hydroxide. Ethyl alcohol will be added to serve as a common solvent for the reactants. You will also investigate the characteristics of soaps formed from fatty acids and some divalent and trivalent cations. These cations are commonly encountered in areas where there is hard water and in industry. Their soaps are usually referred to as **metallic soaps.**

SAFETY PRECAUTIONS AND DISPOSAL METHODS

Alcohol vapors are highly flammable. Keep the flame away from the top of the beaker. Have an asbestos square handy to cover the beaker if the vapors should ignite. Also, sodium hydroxide causes severe burns when in contact with the skin. If a slippery feeling is observed on the hands or other parts of the body, rinse with large amounts of cold water immediately. Wear eye protection at all times during this experiment.

Dispose of excess vegetable oil and ethyl alcohol in separate bottles. Excess NaOH should be diluted with water and then poured down the drain. Collect solutions containing kerosene.

PROCEDURE

1. Preparation of a Soap

A. Weigh a 150-mL beaker on a decigram balance and weigh into it 18 to 20 g of vegetable oil. Add 20 mL of ethyl alcohol and 25 mL of 20% sodium hydroxide solution.

B. Stir the solution and support the beaker on an asbestos gauze on a tripod. Heat the beaker and its contents gently. Continue the heating until the odor of alcohol is no longer apparent and a pasty mass remains in the beaker. The reaction product is a mixture of the soap and the glycerol freed in the reaction (see Equation 32.2).

C. Allow the soap mixture to cool; then add 100 mL of saturated sodium chloride solution and stir thoroughly with a glass rod. This process is called *salting out* and is used to remove the soap from water, glycerol, and any excess sodium hydroxide present.

D. After the mixture has been stirred and mixed completely, filter off the soap on a Büchner funnel, using suction, as illustrated in Figure 31.2. Rinse with two 10-mL portions of ice-cold water, drawing this water through the funnel. Allow your soap to dry by spreading it out on a paper towel.

2. Properties and Reactions of Soaps

A. Washing properties. Take a small amount of your soap and wash your hands with it. In soft water, it should lather easily. If any oil is left over, the soap will feel greasy. Describe the washing properties of your soap on the work page. Rinse your hands several times after the test.

B. Basicity. A soap that contains free alkali is harmful to the skin, silk, or wool. To test for the presence of free base, dissolve a small amount of your soap in 5 mL of ethyl alcohol and add two drops of phenolphthalein. If the indicator turns red, free alkali is present. Record your observation.

C. Reaction with multivalent cations. Dissolve about 1 g of your soap in 50 mL of warm water. Pour about 10 mL of soap solution into each of three test tubes. To the first test tube, add 8 or 10 drops of 5% $CaCl_2$; to the second, 8 or 10 drops of 5% $MgCl_2$; and to the third, 8 or 10 drops of 5% $FeCl_3$. Record your observations on the work page. (Does this remind you of the "scum" that forms when you wash in hard water?)

D. Emulsification. Put 5 to 10 drops of kerosene in a test tube containing 8 to 10 mL of water and shake it. An emulsion or suspension of tiny oil droplets in water will form (the solution will look cloudy). Let this solution stand for a few minutes. Prepare another test tube with the same ingredients, but add about 0.5 g of your soap to it before shaking it. Compare the stabilities of the emulsions in the two test tubes. Which emulsion seems to contain smaller droplets? Which emulsion clears up first? Explain. Record your answers on the work page.

Experiment 32

Advance Study Assignment

1. What is a soap?

2. What does the term *saponification* mean?

3. What does the term *emulsion* mean?

4. What do the terms *hydrophilic* and *hydrophobic* mean?

Experiment 32

Work Page

PART 2—Properties and Reactions of Soaps

Step A—Washing Properties

Soap lathers a lot _____, a little _____, not at all _____.

Soap feels oily: yes _____, no _____.

Step B—Basicity

Soap solution + indicator: turns pink _____; remains colorless _____

Step C—Reaction with Multivalent Cations

Cation Added   Observation

Ca^{2+}

Mg^{2+}

Fe^{3+}

Step D—Emulsification

Emulsion containing smaller droplets:

Kerosene in water_____; kerosene + soap in water _____.

Emulsion that clears up first:

Kerosene in water_____; kerosene + soap in water _____.

Explanation:

Experiment 32

Report Sheet

PART 2—Properties and Reactions of Soaps

Step A—Washing Properties

Soap lathers a lot _____, a little _____, not at all _____.

Soap feels oily: yes _____, no _____.

Step B—Basicity

Soap solution + indicator: turns pink _____; remains colorless _____

Step C—Reaction with Multivalent Cations

Cation Added   Observation

   Ca^{2+}

   Mg^{2+}

   Fe^{3+}

Step D—Emulsification

Emulsion containing smaller droplets:

Kerosene in water _____; kerosene + soap in water _____.

Emulsion that clears up first:

Kerosene in water _____; kerosene + soap in water _____.

Explanation:

Carbohydrates

Performance Goals

33–1 Perform tests on various types of carbohydrates.

33–2 Identify an unknown carbohydrate.

CHEMICAL OVERVIEW

Carbohydrates are a class of organic compounds composed of carbon, hydrogen, and oxygen. Carbohydrates include polyhydroxyaldehydes (aldoses) and polyhydroxyketones (ketoses) or more complex molecules that can be broken down to yield these compounds. Many of our most common foods, such as rice, bread, potatoes, and fruits, are rich in carbohydrates.

The formulas of carbohydrate molecules can be represented by open-chain structures or by ring structures (Figure 33.1). All simple sugars in their open-chain form contain either an aldehyde group, —CHO, or a carbonyl group, $\diagup C{=}O$. Glucose contains six carbons and has a —CHO group in its open-chain form; therefore, it is classified as an **aldohexose.** Fructose is a **ketose** since it contains a ketone (carbonyl) group. Because fructose has six carbon atoms it is referred to as a **ketohexose.** The predominant forms of glucose and fructose in solution are the ring structures.

Figure 33.1
Open-chain and ring structures of glucose

D-glucose

α-D-glucopyranose

419

Further classification of carbohydrates is based on the number of simplest molecules (**monosaccharides**) present. **Disaccharides,** so-called "double sugars," are the result of the combination of two monosaccharides. Common examples of disaccharides are lactose (milk sugar), sucrose (table sugar), and maltose. **Polysaccharides,** such as starch and cellulose, are very large molar mass polymers composed of many monosaccharide units.

Disaccharides and polysaccharides can be broken down to smaller units by hydrolysis (reaction with water). Sucrose and starch can be hydrolyzed under strongly acidic conditions or by using a biological catalyst called an *enzyme*.

Many spot tests are available that allow identification of major carbohydrates. In this experiment we will be using the following tests:

1. **Fehling's test.** This test is used to determine if sugars containing free aldehyde or ketone groups are present. These sugars, called **reducing sugars,** can react with mild oxidizing agents, such as Cu^{2+} in Fehling's solution, to yield Cu_2O, a red-orange solid.

$$Cu^{2+} + \text{reducing sugar} \rightarrow Cu_2O(s) + \text{oxidized sugar}$$
$$(\textit{Red-orange})$$

2. **Iodine test.** Polysaccharides, such as starch, produce a characteristic blue color when they react with a solution containing iodine. It is believed that a complex is formed that varies in intensity and shade of color depending on the size of the molecule, the concentration of iodine, and the temperature.

3. **Seliwanoff's test.** This test is used to differentiate between an aldohexose and a ketohexose. The *quick* appearance of a deep red color is a positive indication of a ketohexose, such as fructose. Aldohexoses react much more slowly.

4. **Barfoed's test.** The difference in reaction rate between reducing monosaccharides and reducing disaccharides with cupric acetate solution is the basis for this test. The smaller the sugar, the faster the reaction rate.

SAFETY PRECAUTIONS AND DISPOSAL METHODS

The reagents used in this experiment contain fairly concentrated acids, bases, and organic compounds that may cause burns or irritation to skin. Avoid contact with all solutions and wear eye protection while performing the experiment. Also, be careful when handling the boiling-water bath.

Collect all your solutions in a waste container or as directed by your instructor.

PROCEDURE

A boiling-water bath is required for some of the tests in this experiment. Pour about 200 mL of deionized water into a 400-mL beaker and heat it to boiling. Maintain it at that temperature, replenishing the water from time to time as it becomes necessary.

1. Fehling's Test

Prepare the reagent for this test by mixing 20 mL of Fehling's solution A with 20 mL of Fehling's solution B. Add 1 mL of 1% solution of glucose, fructose, sucrose, lactose, and starch to separate, labeled test tubes, and then add 5 mL of the mixed Fehling's reagent to each test tube. Heat the test tubes in the boiling-water bath for 5 minutes and record your observations on the work page.

2. Iodine Test

Add 1 mL of 1% solution of glucose, fructose, sucrose, lactose, and starch to separate, labeled test tubes, and then add 3 drops of dilute iodine solution to each. Prepare a sixth test tube with 1 mL of deionized water and 3 drops of iodine solution. Compare the colors observed and record them in the report sheet.

3. Seliwanoff's Test

Dilute the fructose and glucose solutions available to 0.5% (2 mL sugar solution + 2 mL deionized water). Add 1 mL of each into separate test tubes, prepare a third one with 1 mL of deionized water. Add 10 mL of Seliwanoff's reagent to each test tube and place it into a boiling-water bath. Using a stopwatch (or second hand on your watch), record the time required for a color change to occur. Use the deionized water as a blank. Discontinue the heating after 10 minutes.

4. Barfoed's Test

Add 2 mL of 1% solution of fructose, glucose, and lactose to three separate, labeled test tubes. Add 5 mL of Barfoed's reagent to each test tube, mix, and place the test tubes into a boiling-water bath. Observe the time when a reaction becomes apparent (solution becomes cloudy or changes color). Discontinue heating after 15 minutes. Record your observations.

5. Analysis of an Unknown

Obtain an unknown solution from your instructor that may contain one of the following carbohydrates: fructose, glucose, sucrose, lactose, or starch. Using the tests you performed above, determine which compound is present in your solution. Remember, fructose and glucose are monosaccharides, sucrose is a nonreducing disaccharide, lactose is a reducing disaccharide, and starch is a polysaccharide. Be sure to record your unknown number!

Experiment 33

Advance Study Assignment

1. Complete the table below:

Substance	Fehling's Test		Iodine Test		Seliwanoff's Test		Barfoed's Test	
	Yes	No	Yes	No	Yes	No	Fast	Slow
Glucose								
Fructose								
Sucrose					X	X		
Lactose					X	X		
Starch					X	X	X	X

2. What group of compounds gives a positive iodine test? Give an example.

3. (Optional) Sucrose is a disaccharide of glucose and fructose. Honey is a mixture of glucose and fructose. Explain why honey is a quicker energy source than table sugar (sucrose).

Experiment 33

Work Page

Part 1—Fehling's Test

Substance	Test Tube	Yes	No
Glucose	1		
Fructose	2		
Sucrose	3		
Lactose	4		
Starch	5		
Unknown			

Part 2—Iodine Test

Substance	Test Tube	Color
Glucose	1	
Fructose	2	
Sucrose	3	
Lactose	4	
Starch	5	
H_2O	6	
Unknown		

Part 3—Seliwanoff's Test

Substance	Test Tube	Yes	No	Time
Fructose	1			
Glucose	2			
H_2O	3			
Unknown				

Part 4—Barfoed's Test

Substance	Test Tube	Time
Fructose	1	
Glucose	2	
Lactose	3	
Unknown		

Part 5—Analysis of an Unknown

Unknown number _____   Compound present _____

Experiment 33

Report Sheet

Part 1—Fehling's Test

Substance	Test Tube	Yes	No
Glucose	1		
Fructose	2		
Sucrose	3		
Lactose	4		
Starch	5		
Unknown			

Part 2—Iodine Test

Substance	Test Tube	Color
Glucose	1	
Fructose	2	
Sucrose	3	
Lactose	4	
Starch	5	
H_2O	6	
Unknown		

Part 3—Seliwanoff's Test

Substance	Test Tube	Yes	No	Time
Fructose	1			
Glucose	2			
H_2O	3			
Unknown				

Part 4—Barfoed's Test

Substance	Test Tube	Time
Fructose	1	
Glucose	2	
Lactose	3	
Unknown		

Part 5—Analysis of an Unknown

Unknown number _____   Compound present _____

Lipids

Performance Goals

34–1 Determine whether a lipid is saturated or unsaturated.

34–2 Identify an unknown as stearic acid, olive oil, lecithin, or cholesterol.

CHEMICAL OVERVIEW

The word *lipid* comes from the Greek *lipos*, which means "fat." Fats or lipids are defined as molecules that come from cells or tissues that are soluble in nonpolar solvents. Due to the definition being based mainly on solubility, lipids have a large range of structural diversity. This structural range can be further divided into four major categories: fatty acids (saturated and unsaturated), glycerides (glycerol-containing), nonglycerides (sphingolipids, steroids, waxes), and complex lipids (lipoproteins). In this experiment, three glycerol-containing lipids will be explored: a fat or oil, which contains glycerol and three fatty acids (saturated and unsaturated); lecithin, which contains glycerol; and two fatty acids and a phosphate attached to an amino alcohol. Cholesterol represents the steroid, nonglyceride lipids.

Fatty acids are carboxylic acids attached to long hydrocarbon chains of 10 to 20 carbons. Each fatty acid is further characterized by the length of the carbon chain and by the number of double bonds. Fatty acids with double bonds are unsaturated fatty acids, and those without double bonds are saturated fatty acids. The term *polyunsaturated fat* refers to a fatty acid that has multiple double bonds.

A qualitative test to determine saturation or unsaturation of fatty acids is to react the acid with a bromine solution in methlyene chloride. Bromine readily reacts with any alkene to dihalogenate. The bromine solution is reddish-brown in color and will become clear as the bromine reacts with any alkene present.

Saturated Fatty Acids	*Unsaturated Fatty Acids*
Animal fat, stearic acid	Oleic acid (contained in olive oil)
$CH_3(CH_2)_{16}COOH$	$CH_3(CH_2)_7CH=CH(CH_2)_7COOH$
	Linoleic acid (contained in safflower oil)
	$CH_3(CH_2)_4CH=CHCH_2CH=CH(CH_2)_7COOH$

Glycerides are lipids that contain the molecule glycerol and fatty acids. Triglycerides, which contain one glycerol molecule and three fatty acids, are the main constituents of oils and animal fats. Fats tend to contain saturated fatty acids and come from animals, while oils contain unsaturated fatty acids and usually come from plants. A typical triglyceride found in olive oil consists of two oleic fatty acids and one palmitic fatty acid attached by an ester bond to glycerol. In this experiment, triglycerides containing unsaturated fatty acids will be identified by the bromine test.

$$
\begin{array}{l}
H_2C{-}O{-}\overset{\overset{\displaystyle O}{\|}}{C}{-}(CH_2)_7CH{=}CH(CH_2)_7CH_3 \\[2pt]
CH{-}O{-}\overset{\overset{\displaystyle O}{\|}}{C}{-}(CH_2)_7CH{=}CH(CH_2)_7CH_3 \quad + \quad Br_2 \\[2pt]
H_2C{-}O{-}\overset{\overset{\displaystyle O}{\|}}{C}{-}(CH_2)_{14}CH_3
\end{array}
\quad\longrightarrow\quad
\begin{array}{l}
H_2C{-}O{-}\overset{\overset{\displaystyle O}{\|}}{C}{-}(CH_2)_7\overset{Br}{CH}{-}\overset{Br}{CH}(CH_2)_7CH_3 \\[2pt]
CH{-}O{-}\overset{\overset{\displaystyle O}{\|}}{C}{-}(CH_2)_7\overset{Br}{CH}{-}\overset{Br}{CH}(CH_2)_7CH_3 \\[2pt]
H_2C{-}O{-}\overset{\overset{\displaystyle O}{\|}}{C}{-}(CH_2)_{16}CH_3
\end{array}
$$

Olive oil   (orange color)   (colorless)   (34.1)

Another variation of triglycerides is phosphoglycerides, which are compounds that contain a phosphoryl group. When it is attached to a choline molecule, this lipid is designated as a lecithin.

A test for the presence of a phosphate can be used indirectly to differentiate between a phospholipid, cholesterol, and fat. Lecithin can be hydrolyzed in an acidic medium to give its four major constituents: glycerol, fatty acids, choline, and phosphate. The phosphate can then be detected using the molybdate test in which the appearance of a light-yellow precipitate indicates a positive test.

$$
\begin{array}{l}
H_2C{-}O{-}\overset{\overset{\displaystyle O}{\|}}{C}{-}(CH_2)_nCH_3 \\[2pt]
CH{-}O{-}\overset{\overset{\displaystyle O}{\|}}{C}{-}(CH_2)_nCH_3 \quad + \quad 4H_2O \\[2pt]
H_2C{-}O{-}\overset{\overset{\displaystyle O}{\|}}{P}{-}O{-}CH_2CH_2\overset{+}{N}{\overset{\textstyle CH_3}{\underset{\textstyle CH_3}{\diagup}}}{-}CH_3 \\[2pt]
\underset{\textstyle O^-}{}
\end{array}
\overset{H^+}{\rightleftharpoons}
\begin{array}{l}
CH_2OH \\ | \\ CHOH \\ | \\ CH_2OH \\ \text{glycerol}
\end{array}
\;+\;
\begin{array}{l}
O \\ \| \\ 2\,HO{-}C{-}(CH_2)_nCH_3\;+ \\ \text{fatty acid}
\end{array}
$$

lecithin

$$
HO{-}CH_2CH_2\overset{+}{N}{\overset{\textstyle CH_3}{\underset{\textstyle CH_3}{\diagup}}}{-}CH_3
\quad + \quad
HO{-}\overset{\overset{\displaystyle O}{\|}}{\underset{\displaystyle O^-}{P}}{-}OH
$$

choline   phosphate   (34.2)

Steroids are classified as nonglycerol lipids that have a fused, four-ring skeleton system. Each ring is designated with a letter as shown below. An example of a common steroid is cholesterol.

The Lieberman–Buchard test is used to determine the presence of cholesterol. The reagent contains acetic anhydride and sulfuric acid, which reacts with the alcohol (OH group) and the alkene (double bond) in the adjacent ring. A gradual color change from pink to lilac and then to deep green will be observed in a positive test.

cholesterol

SAFETY PRECAUTIONS AND DISPOSAL METHODS

Acids and bases are corrosive, and contact with your skin should be avoided. Any spilled acid or base should be washed off promptly. Sulfuric acid will burn when in contact with skin; wipe off the acid, then rinse the exposed area with plenty of cold water. Be sure to wear safety goggles or safety glasses while performing this experiment. Bromine, Br_2, causes severe burns on contact with the skin. Handle this reagent very carefully. Organic solvents are volatile and inhaling the vapors should be avoided.

Dispose of excess solids and liquids as directed by your instructor. Solutions containing precipitates should be collected in a waste container.

PROCEDURE

1. Bromine Test for Unsaturation

A. Prepare four test tubes by placing 5 drops, or a small amount of solid, of the following into separate test tubes: stearic acid, olive oil, lecithin, and cholesterol.

B. Add 1 mL of methlyene chloride to each of the test tubes. Then, add 2% Br_2 in methylene chloride to each sample drop by drop until a permanent orange-red color is obtained, or until 20 drops have been added. Record the final color.

2. Lieberman–Buchard Test for Cholesterol

A. Prepare four test tubes by placing 5 drops, or a small amount of solid, of the following into separate test tubes: stearic acid, olive oil, lecithin, and cholesterol. Add 3 mL of chloroform and 1 mL of acetic anhydride to each test tube. Next, add 1 drop of concentrated sulfuric acid to each mixture and record the color. Set aside for 5 minutes and record the color of each test tube.

3. Phosphate Test

A. Prepare four test tubes by placing 5 drops, or a small amount of solid, of the following into separate test tubes: stearic acid, olive oil, lecithin, and cholesterol.

B. Hydrolyze the compounds by adding 3 mL of 6 M nitric acid to each test tube.

C. Place each tube in a boiling hot water bath for 5 minutes. Remove the test tubes and let them cool to room temperature. Neutralize the samples by adding 3 mL of 6 M NaOH to each test tube. If a precipitate forms, the sample needs to be filtered.

D. Add 1 mL of molybdate solution to each test tube. Mix thoroughly and heat in a hot-water bath for 5 minutes. A light-yellow precipitate indicates that phosphate is present.

CAUTION

You need to have a *precipitate*; a yellow color is not a positive test.

4. Identification of an Unknown

Obtain an unknown from your instructor and perform tests 1 through 3 to determine the identity of your lipid as stearic acid, olive oil, lecithin, or cholesterol. Be sure to record the unknown number.

Experiment 34

Advance Study Assignment

1. Write the equation for the reaction of linoleic acid with bromine.

2. a. What is the purpose of adding acid to the lecithin sample?

    b. Write the four products of this reaction (from part 3b).

    c. Which of the four products in part 3c is responsible for a positive test?

3. Identify the two unknowns below as stearic acid, olive oil, lecithin, or cholesterol, based on the following results.

    a. The unknown reacts to give a green color in the presence of acetic anhydride and sulfuric acid.

    b. The unknown gives negative results for tests 1 through 3.

Experiment 34

Work Page

A. Bromine Test for Unsaturation

	Number of drops added	Ending color	Saturated	Unsaturated
Stearic acid				
Olive oil				
Lecithin				
Cholesterol				
Unknown				

B. Lieberman–Buchard Test for Cholesterol

	Initial color	Ending color	Cholesterol present
Stearic acid			
Olive oil			
Lecithin			
Cholesterol			
Unknown			

C. Phosphate Test for Lecithin

	Precipitate	*Phosphate present*
Stearic acid		
Olive oil		
Lecithin		
Cholesterol		
Unknown		

Unknown number _____ Identity of unknown _____

Experiment 34

Report Sheet

A. Bromine Test for Unsaturation

	Number of drops added	Ending color	Saturated	Unsaturated
Stearic acid				
Olive oil				
Lecithin				
Cholesterol				
Unknown				

B. Lieberman–Buchard Test for Cholesterol

	Initial color	Ending color	Cholesterol present
Stearic acid			
Olive oil			
Lecithin			
Cholesterol			
Unknown			

C. Phosphate Test for Lecithin

	Precipitate	Phosphate present
Stearic acid		
Olive oil		
Lecithin		
Cholesterol		
Unknown		

Unknown number _____ Identity of unknown _____

Amino Acids and Proteins

Performance Goal

35–1 Perform identification tests on amino acids and proteins.

CHEMICAL OVERVIEW

Amino acids are molecules containing two functional groups: the amino group ($-NH_2$) and the carboxyl group ($-COOH$). Proteins are large, complex molecules, built from amino acids that are joined by peptide linkages. A peptide bond is formed when an acidic carboxyl group of one amino acid reacts with the basic amino group of another amino acid molecule. The formation of such a bond is shown in Equation 35.1.

$$R_1-CH-C + H_2N-CH-C \xrightarrow{-H_2O} R_1-CH-C-NH-CH-C \quad (35.1)$$

Peptide bond

The product still contains a free amino group, which gives basic properties to the molecule, and a free carboxyl group, which gives it acidic properties.

When two amino acids react with each other, the product is a **dipeptide.** When a very large number of amino acids are linked, the **polypeptides** that form are known as **proteins.** These molecules make up our skin, muscles, and enzymes. Hormones, hair, and fingernails are also made of protein. In order for these molecules to function, they must exist in a specific, three-dimensional structure. Biological activity will cease if this structure is destroyed. Extreme temperatures, acids, bases, and heavy metal compounds can break linkages in proteins, causing them to become **denatured.**

In this experiment you will carry out some tests that are characteristic of amino acids and will investigate the various means of coagulating (denaturing) proteins.

1. **Biuret test.** This test identifies a compound that contains two or more peptide bonds. When these compounds react with Cu^{2+} ions in a basic solution, a pink-violet complex is formed (a positive test). Amino acids

and dipeptides (two amino acids joined by one peptide bond) do not give violet colors with Cu^{2+}, but produce a blue solution (a negative test).

2. **Xanthoproteic test.** Some amino acids and the proteins containing them have aromatic rings. These rings react with concentrated nitric acid to produce a yellow compound, which is intensified in a basic solution. The test is used to determine the presence or absence of aromatic ring structures.

3. **Hopkins–Cole test.** This is a specific test for the presence of proteins containing tryptophan, one of the essential amino acids. Eggs, for example, are high in tryptophan. When a tryptophan-containing protein solution is layered on top of concentrated sulfuric acid, a purple compound (ring) forms at the interface.

4. **Unoxidized sulfur test.** This test is to detect the presence of sulfur-containing amino acids (such as cysteine) or proteins containing these amino acids. In a basic solution inorganic sulfide ions, S^{2-}, are produced, which react with lead acetate to give PbS, a black precipitate.

SAFETY PRECAUTIONS AND DISPOSAL METHODS

The reagents used in this experiment contain organic compounds that may irritate skin. Concentrated nitric and sulfuric acids are *very corrosive*. Be sure not to get them on your skin or clothing. If you have spilled concentrated sulfuric acid on your skin, *wipe it off first*, then wash with soap and water. *WEAR EYE PROTECTION THROUGHOUT THE EXPERIMENT.* Also, handle the hot-water bath carefully.

Discard solutions containing heavy metal ions in a stoppered bottle.

PROCEDURE

A boiling-water bath is required for some of the tests in this experiment. Pour about 200 mL of deionized water into a 400-mL beaker and heat it to boiling. Maintain it at that temperature, replenishing the water as it becomes necessary.

1. Biuret Test

Label four clean test tubes. Pour 2 mL of egg albumin solution, 1% gelatin solution, 1% casein solution, and 0.5% alanine solution (an amino acid) into separate test tubes. Now add 3 mL of 10% sodium hydroxide solution to each one and shake carefully to mix. Add 2 drops of 2% copper sulfate solution to each test tube, mix, and observe the color of the solution. Record your observations on the work page.

2. Xanthoproteic Test

Pour 2 mL of egg albumin solution into a test tube and add 10 drops of concentrated nitric acid to it. Mix and heat in a boiling-water bath for 2 minutes. Record any color change. Cool the mixture and add 10% sodium hydroxide solution dropwise until the mixture is basic to litmus (red litmus turns blue). Note any change in color. Record your observations.

3. Hopkins–Cole Test

Label three clean test tubes. Pour 2 mL of egg albumin solution, 1% tryptophan solution, and 1% gelatin solution into separate test tubes. Add to each one 2 mL of Hopkins–Cole reagent and mix. Now, holding the first test tube at a 45° angle, *VERY CAREFULLY AND SLOWLY* add 30 drops of concentrated sulfuric acid down the inside wall of the test tube, so that it forms a layer on the bottom. Avoid shaking the tube or doing anything that will cause mixing. The presence of a purple ring at the interface is a positive test for tryptophan. If no ring forms, gently tap the test tube to cause slight mixing at the interface. Record your observations on the work page. Repeat the procedure for the second and third test tubes.

4. Unoxidized Sulfur Test

Pour 2 mL of egg albumin solution into a test tube. Add 5 mL of 10% sodium hydroxide solution and 3 drops of 5% lead acetate solution. Mix and heat in a boiling-water bath for 3 minutes. Observe the results and record them on the work page.

5. Coagulation/ Precipitation of Proteins

Obtain five test tubes and pour 3 mL of egg albumin solution into each one.

A. Heat the first test tube to boiling. Observe any changes and record your results on the work page.

B. To the second test tube add 10 mL of 95% ethanol (ethyl alcohol). Mix and observe changes. Record your observations on the work page.

C. To the third test tube add 2% silver nitrate solution dropwise. Record your observations.

D. Using the fourth test tube, add 5% mercury(II) chloride, $MgCl_2$, dropwise until you see a change. Record your observations.

E. To the fifth test tube, add dilute tannic acid dropwise until you see a change. Record your observations.

Experiment 35

Advance Study Assignment

1. Which two functional groups are present in all amino acids?

2. Explain why your skin turns yellow if it comes in contact with concentrated nitric acid.

3. What process takes place when milk goes sour?

4. (Optional) Why is alcohol used as a disinfectant before you get a shot?

5. (Optional) When a person is suspected to have ingested a heavy metal salt, such as $AgNO_3$ or $Pb(NO_3)_2$, why is egg white or milk used as an antidote?

Experiment 35

Work Page

Part 1—Biuret Test

Test Tube	Substance	Yes	No
1	Egg albumin		
2	Gelatin		
3	Casein		
4	Alanine		

If you heated a protein solution with an acid to break the bonds and to produce the amino acid "building blocks," would you expect the hydrolysis product to give a positive _____ or negative _____ Biuret test? Explain.

Part 2—Xanthoproteic Test

Original color of solution: _____

Color after heating with HNO_3: _____

Color of basic solution: _____

Part 3—Hopkins–Cole Test

Test Tube	Substance	Observation
1	Egg albumin	
2	Triptophan	
3	Gelatin	

Part 4—Unoxidized Sulfur Test

Formation of a precipitate: Yes _____ ;   No _____

Write the equation for the reaction, if any.

Part 5—Coagulation/Precipitation of Proteins

Step	Observation
5A	
5B	
5C	
5D	
5E	

Questions (Optional)

1. What happens to egg albumin when an egg is hard boiled?

2. A dilute solution of mercury(II) chloride can be used to preserve anatomical specimens. Explain why.

3. What happens chemically to proteins in your body during digestion?

Experiment 35

Report Sheet

Part 1—Biuret Test

Test Tube	Substance	Yes	No
1	Egg albumin		
2	Gelatin		
3	Casein		
4	Alanine		

If you heated a protein solution with an acid to break the bonds and to produce the amino acid "building blocks," would you expect the hydrolysis product to give a positive _____ or negative _____ Biuret test? Explain.

Part 2—Xanthoproteic Test

Original color of solution: _____

Color after heating with HNO_3: _____

Color of basic solution: _____

Part 3—Hopkins–Cole Test

Test Tube	Substance	Observation
1	Egg albumin	
2	Triptophan	
3	Gelatin	

Part 4—Unoxidized Sulfur Test

Formation of a precipitate: Yes _____ ;   No _____

Write the equation for the reaction, if any.

Part 5—Coagulation/Precipitation of Proteins

Step	Observation
5A	
5B	
5C	
5D	
5E	

Questions (Optional)

1. What happens to egg albumin when an egg is hard boiled?

2. A dilute solution of mercury(II) chloride can be used to preserve anatomical specimens. Explain why.

3. What happens chemically to proteins in your body during digestion?

Enzymes

Performance Goals

36–1 Determine the activity of an enzyme.

36–2 Determine how the rate of an enzyme-catalyzed reaction changes under different conditions.

CHEMICAL OVERVIEW

An enzyme is a biological molecule that increases the rate of a reaction by lowering the activation energy. This allows reactions to occur approximately a million times faster and at much lower temperatures. Without increasing the rate of most biochemical reactions, life would not be possible.

Nearly all known biological catalysts are proteins that function by folding into a specific three-dimensional form called the tertiary structure. This structure is primarily due to four intermolecular forces: hydrogen bonding, nonpolar interactions, ionic bonds, and disulfide linkages. The enzyme's ability to function will be diminished or stopped if these intermolecular forces are disrupted, which, in turn, will alter the tertiary structure. This process of diminishing, or stopping the enzyme's function by altering its shape, is called denaturing. There are two major ways in which an enzyme can be denatured: alter its environment or introduce an enzyme inhibitor.

Most enzymes function in a specific biological environment. Some common causes of denaturing include heat and pH changes. In this experiment, both of these factors will be used to make the enzymes inoperative.

Silver nitrate will be used as an inhibitor. Heavy metals, like silver, often bind to the sulfhydryl groups of the amino acid cysteine to reduce or stop enzyme activity. Silver nitrate was used in newborn infants' eyes as an antibacterial agent.

This experiment will use catalase as an enzyme to work on the substrate hydrogen peroxide to form water and oxygen. Hydrogen peroxide is toxic to cells, so this enzyme's function is to decrease peroxide concentrations. Catalase is found in red blood cells and in many plants.

$$2H_2O_2 \xrightarrow{\text{catalase}} 2H_2O(l) + O_2(g) \qquad (36.1)$$

The rate of the reaction will be measured by collecting and measuring the amount of oxygen generated in 5 minutes. Oxygen will be collected in a graduated cylinder initially full of water. The amount of water displaced will equal the volume of oxygen gas generated from the enzyme. The time of the reaction will be measured, and the rate will then be reported as milliliters per minute.

Volume of $O_2 = V_{final} - V_{initial}$

Rate of catalase (mL/min) = Volume of O_2/5 minutes

SAFETY PRECAUTIONS AND DISPOSAL METHODS

Be sure to wear safety goggles or safety glasses while performing this experiment. Acids are corrosive and contact with the skin should be avoided. Any spilled acid should be washed off promptly.

Dispose of the waste as directed by your instructor.

PROCEDURE

NOTE: Record all volume measurements to the nearest 0.1 mL.

1. Standardization of Enzyme Activity

A. Put about 400 mL of water into a 500-mL beaker. Next, take a 250-mL beaker and fill it near the top with water and place on a hot plate.

B. Set up the apparatus according to Figure 36.1. Heat the water bath to 35–37°C. Often when the graduated cylinder is tipped over, a small amount of air will be trapped at the top. Record this initial amount of air in the 25-mL graduated cylinder.

C. Using a graduated, cylinder measure 5.0-mL of an 18% solution of H_2O_2. Place a magnetic stirring bar into a test tube and then pour in the solution. Place the stopper in the test tube.

D. Open the stopper and add 5.0 mL of the catalase solution. Immediately stopper the test tube and begin recording the time. This step must be completed quickly so no oxygen is lost. Shake the test tube to insure mixing. Turn on the magnetic stirrer.

Figure 36.1
Enzyme rate measuring apparatus

E. Wait until 5 minutes have elapsed. Record the final volume of oxygen in the graduated cylinder.

F. Repeat steps B through E. Record both results and average them.

2. Double Substrate Concentration

A. Follow the directions outlined above except for step 1C. Instead, pour 10.0 mL of the catalase solution into the test tube and stopper.

3. Double Enzyme Concentration

A. Follow the directions outlined above except for step 1D. Instead, pour 10.0 mL of an 18% solution of H_2O_2 into the test tube and stopper.

4. Heat Effect

A. Place a test tube containing 5.0 mL of the catalase solution into a boiling-water bath for 5 minutes. Remove and let it cool.

B. Pour 5.0 mL of an 18% solution of H_2O_2 into another test tube and put it in the water bath at 35–37°C and stopper it. Record the initial amount of air in the graduated cylinder.

C. Open the stopper and add 5.0 mL of the catalase that had previously been heated in part 4A. Immediately stopper the test tube and begin recording the time. This step must be completed quickly so no oxygen is lost.

D. Wait until 5 minutes have elapsed. Record the final volume of oxygen in the graduated cylinder.

5. Acidic Conditions

A. Add 5 drops of 0.1 M HCl to a test tube containing 5.0 mL of the catalase solution.

B. Measure the activity of the enzyme by following the procedure above (part 1, steps B through E).

6. Inhibitor

A. Add 5 drops of 0.1 M $AgNO_3$ solution to a test tube containing 5.0 mL of the catalase solution.

B. Measure the activity of the enzyme by following the procedure above (part 1, steps B through E).

Experiment 36

Advance Study Assignment

1. Calculate the rate of catalase if the final volume after 10 minutes is 34.5 mL. The initial volume was 4.2 mL.

2. Identify the substrate in the experiment.

3. What enzyme rate is expected for the boiled catalase?

4. What happens to a protein when it is denatured?

5. What is an enzyme and what is its function?

Experiment 36

Work Page

Results

	Water bath Run 1	Water bath Run 2	Double sub-strate conc.	Double enzyme conc.	Boiled catalase	Acidic condition	Inhibitor AgNO$_3$
Final volume (mL, 5 min)							
Initial volume (mL)							
Volume of O$_2$ (mL)							
Rate of enzyme (mL/min)							

Average of rate for part 1 (water bath at 35–37°C). _____

Show all your calculations below:

Questions

1. Which conditions resulted in the largest enzyme rate?

2. Did boiling the enzyme and using an inhibitor decrease the rate of the enzyme to the same degree?

3. What rate was measured for the boiled catalase? Explain this result.

4. What rate was measured for the acidic catalase? Explain this result.

5. What rate was measured when the substrate concentration was doubled? Was this new rate expected?

Experiment 36

Report Sheet

Results

	Water bath Run 1	Water bath Run 2	Double substrate conc.	Double enzyme conc.	Boiled catalase	Acidic condition	Inhibitor AgNO$_3$
Final volume (mL, 5 min)							
Initial volume (mL)							
Volume of O$_2$ (mL)							
Rate of enzyme (mL/min)							

Average of rate for part 1 (water bath at 35–37°C). _____

Show all your calculations below:

Questions

1. Which conditions resulted in the largest enzyme rate?

2. Did boiling the enzyme and using an inhibitor decrease the rate of the enzyme to the same degree?

3. What rate was measured for the boiled catalase? Explain this result.

4. What rate was measured for the acidic catalase? Explain this result.

5. What rate was measured when the substrate concentration was doubled? Was this new rate expected?

Worksheets

SIGNIFICANT FIGURES

1. Rewrite the following numbers using scientific notation:

   a. 476

   b. 0.00367

   c. 549×10^3

   d. 0.0000069

   e. 9546.3

   f. 0.00485×10^7

   g. 264×10^{-5}

   h. 4795

   i. 67.09

   j. 100.

2. How many significant figures are there in each of the following numbers?

   a. 16.0

   b. 54,056

   c. 1000

   d. 0.00594

   e. 207.3

   f. 10

   g. 5.2×10^7

   h. 1.68×10^{-9}

   i. 2007

   j. 5×10^2

3. Round the following numbers to three significant figures and use scientific notation where appropriate:

   a. 7894

   b. 0.00003982

   c. 100378

   d. 19047×10^{-2}

   e. 2345

   f. 3.075

4. Perform the following mathematical operations and express your answers to the proper number of significant figures:

   a. $645 \times 2.0 \times 167.8$

   b. $0.045 \times 128.2 \times 34.6$

   c. $190.4 + 12 + 0.69$

   d. $26.6 \times (3.7 \times 10^2)$

   e. $(3.65 \times 10^4) \times (2.1 \times 10^2)$

   f. $(4.3 \times 10^3)^5$

   g. $2597/42$

   h. $12.0/1.8 \times 10^{23}$

   i. $3.006/4.68 \times 10^{-4}$

   j. $1.2 + 45.81 + 0.186$

DIMENSIONAL ANALYSIS

Set up and solve the following problems using dimensional analysis. Be sure to express your results to the proper number of significant figures.

1. How many seconds are there in 1.2 weeks?
2. How many centimeters are there in 4.38 feet?
3. How many meters did you run if the distance run was 6.59×10^5 inches?
4. What is the mass of a suitcase, in pounds, if it weighs 19.5 kilograms?
5. If a recipe calls for 37 grams of sugar, how many pounds does that correspond to?
6. Express a volume of 589 cm^3 in ft^3 and in^3.
7. How many liters are equal to 39 in^3?
8. If a car travels at 4.45×10^4 ft/hr, what would its speed be in meters/min?
9. What is the density of a substance if it has a mass of 59.2 grams and a volume of 17.0 mL?
10. Calculate the density of a liquid, in grams/cm^3, if it has a mass of 23.2 grams and occupies a container with dimensions of 1.3 cm × 5.6 cm × 2.3 in.
11. If a liquid has a density of 1.04 g/mL, what would its density be in lb/in^3?
12. What volume, in liters, would 88.9 grams of a substance occupy if its density is 2.38 g/mL?
13. What is the mass, in pounds, of 389 mL of a gas that has a density of 1.29 g/L?
14. Convert 37°C to °F and K.
15. Which temperature is the coldest?

    a. −12°C   b. 18°F   c. 248K

AVOGADRO'S NUMBER MOLES

1. What is the molar/formula mass of the following species?

   a. H_2SO_4

   b. Na_2CO_3

   c. CO_2

   d. K_3PO_4

   e. $FeCl_2$

   f. carbon tetra bromide

   g. aluminum hydroxide

   h. copper(II) iodide

   i. tin (IV) sulfate

   j. magnesium oxide

2. How many moles are represented by each of the following amounts?

   a. 4.36 g of nickel

   b. 20.5 g of hydrochloric acid

   c. 205.3 g of calcium hydroxide

   d. 47 g of sulfur dioxide

   e. 3.60 g of sodium sulfide

   f. 0.00874 g of dinitrogen tetroxide

3. What is the mass, in grams, of the following species?

   a. 3.5 moles of silver phosphate

   b. 0.00874 moles of zinc carbonate

   c. 9.0 moles of iron(III) chloride

   d. 1.59 moles of lead(II) oxide

   e. 4.60 moles of sulfurous acid

   f. 0.0695 moles of potassium hydrogen sulfide

4. Answer the following questions:

   a. How many atoms are there in 3.5 grams of copper?

   b. How many atoms are there in 0.143 gram of aluminum?

   c. How many molecules are there in 54.3 grams of barium hydroxide?

   d. What is the mass of 1.15×10^{20} molecules of lithium sulfate?

   e. How many moles are represented by 3.6×10^{34} molecules of water?

   f. 58.0 grams of sodium oxide correspond to how many molecules?

   g. Which amount contains the largest number of molecules?

   a. 2.67 g of water

   b. 9.45 g of sulfuric acid

   c. 10.0 g of sulfur trioxide

FORMULA WRITING

Cation \ Anion	Chloride	Sulfate	Carbonate	Nitrate
Ammonium				
Silver				
Iron(III)				
Sodium				
Potassium				
Copper(II)				

Cation \ Anion	Hydroxide	Oxide	Phosphate	Sulfide
Barium				
Magnesium				
Lead(II)				
Mercury(II)				
Copper(I)				
Cobalt(II)				

EQUATION BALANCING TYPES OF REACTIONS

Using whole-number coefficients only, **balance** the equations below:

a. $NaOH(aq) + H_3PO_4(aq) \rightarrow Na_3PO_4(aq) + H_2O(\ell)$

b. $AgNO_3(aq) + H_2SO_4(aq) \rightarrow Ag_2SO_4(s) + HNO_3(aq)$

c. $Zn(s) + FeCl_3(aq) \rightarrow Fe(s) + ZnCl_2(aq)$

d. $Fe(NO_3)_3(aq) + KOH(aq) \rightarrow Fe(OH)_3(s) + KNO_3(aq)$

e. $CaCl_2(aq) + Al_2(SO_4)_3(aq) \rightarrow CaSO_4(s) + AlCl_3(aq)$

f. $N_2(g) + H_2(g) \rightarrow NH_3(g)$

g. $CH_4(g) + O_2(g) \rightarrow CO_2(g) + H_2O(\ell)$

h. $CH_3OH(\ell) + O_2(g) \rightarrow CO_2(g) + H_2O(\ell)$

i. $KClO_3(s) \rightarrow KCl(s) + O_2(g)$

j. $Fe_2(CO_3)_3(s) + HCl(aq) \rightarrow FeCl_3(aq) + H_2O(\ell) + CO_2(g)$

k. $Cu(NO_3)_2(aq) + H_3PO_4(aq) \rightarrow Cu_3(PO_4)_2(s) + HNO_3(aq)$

OPTIONAL

Classify each of the reactions above as:

1. double replacement
2. double replacement—acid–base reaction (neutralization)
3. double replacement—precipitation
4. single replacement (redox)
5. combination (synthesis)
6. decomposition
7. combustion (complete oxidation)

STOICHIOMETRY

In the following exercises the equations are **unbalanced.** As a first step, be sure to balance the equations and then proceed with the calculations.

1. $NaOH + H_3PO_4 \rightarrow Na_3PO_4 + H_2O$

   a. How many grams of phosphoric acid are required to react with 27.4 g of sodium hydroxide?

   b. How many grams of water are produced from 18.5 g of phosphoric acid?

   c. How many grams of sodium phosphate are produced from 34.7 g of sodium hydroxide?

   d. How many grams of sodium hydroxide are needed to produce 2.56 g of water?

2. $CH_3OH + O_2 \rightarrow H_2O + CO_2$

   a. How many grams of carbon dioxide are produced from the combustion of 8.42 g of methanol?

   b. How many grams of oxygen are required to fully burn 3.59 g of methanol?

   c. How many grams of oxygen will produce 12.6 g of water?

   d. How many grams of methanol will react with 4.65 g of oxygen?

3. $AgNO_3 + AlBr_3 \rightarrow AgBr + Al(NO_3)_3$

   a. How many grams of silver bromide can be produced from 43.6 g of silver nitrate?

   b. How many grams of aluminum bromide are needed to produce 124 g of silver bromide?

   c. How many grams of silver nitrate will produce 74.4 g of aluminum nitrate?

   d. How many grams of aluminum bromide are needed to react with 18.6 g of silver nitrate?

4. $Zn + Pb(NO_3)_4 \rightarrow Pb + Zn(NO_3)_2$

   a. How many grams of zinc will yield 43.9 g of lead?

   b. How many grams of lead(IV) nitrate will produce 1.47 g of zinc nitrate?

   c. How many grams of zinc are needed to react with 4.86 g of lead(IV) nitrate?

   d. 4.58 lbs of zinc will produce how many grams of lead?

ATOMIC STRUCTURE

1. The atomic number of an element equals the number of _____ and _____ in that element.

2. The mass number is the sum of _____ and _____.

3. The atomic mass is the weighted average of _____.

4. Isotopes have the same number of _____ and _____ but a different number of _____.

5. Write the spectroscopic notation ($1s^2 2s^2 2p^6$, etc.) for
   a. aluminum
   b. chromium
   c. Cu^{2+}
   d. Fe^{3+}
   e. arsenic

6. Which species corresponds to each of the following electron structures?
   a. $1s^2 2s^2 2p^6 3s^2 3p^4$ _____
   b. $[Ar]4s^2 3d^3$ _____
   c. $[Ar]3d^7$ _____
   d. $[Kr]5s^2 4d^{10} 5p^2$ _____
   e. $1s^2 2s^2 2p^6 3s^2 3p^6 4s^1$ _____

7. Which of the following contains species that are not isoelectronic?
   a. Ne, F^- b. N^{3-}, O^{2-} c. Ca^{2+}, Mg^{2+} d. Ar, K^+

8. Which of the following set(s) of quantum numbers is/are not possible?

   a. $n = 2$ b. $n = 3$ c. $n = 2$ d. $n = 4$

   $l = 1$ $l = 2$ $l = 2$ $l = 0$

   $m = 0$ $m = -1$ $m = 1$ $m = 1$

   $s = +\frac{1}{2}$ $s = -\frac{1}{2}$ $s = -\frac{1}{2}$ $s = +\frac{1}{2}$

9. Name the orbitals that correspond to the following quantum numbers:
   a. $n = 3$ and $l = 2$ _____
   b. $n = 4$ and $l = 3$ _____
   c. $n = 2$ and $l = 0$ _____
   d. $n = 6$ and $l = 1$ _____

10. Complete the following statements:
    a. There are _____ p orbitals which are _____ to each other.
    b. s orbitals have a _____ shape.
    c. When the first row of transition elements ionize, the _____ electrons are lost before the _____ are lost.

GASES

1. If a gas sample at constant temperature has a volume of 21.3 mL at 764 torr, what will be the volume at 546 torr?

2. A 5.40-L gas sample at 854 torr is compressed at 1.35 atm to what volume?

3. What will be the temperature of a gas sample, in °C, if 2.0 L at 0°C is expanded to 2.5 L? The initial pressure was 750 torr; the final pressure is 785 torr.

4. At constant pressure, a gas sample occupies 379 mL at 23°C. At what temperature will the volume be doubled?

5. 269 mL of a gas sample has a pressure of 1.2 atm at 34°C. Calculate the pressure, in torrs, if the volume is increased to 459 mL and the temperature is decreased to 12°C.

6. What will be the volume of a gas sample containing 12.4 grams of carbon dioxide at STP?

7. If a gas at 25°C and 755 torr has a volume of 333 mL, what would its volume be at STP?

8. How many grams of chlorine are contained in 57.5 mL of gas at 26°C and 786 torr?

9. What pressure, in torrs, will be exerted by 3.57 grams of nitrogen monoxide at 34°C if its volume is 59.0 mL?

10. What is the molar mass of a gas if 286 mL at 69°C and 17.9 torr pressure weighs 0.0156 g?

11. What is the molar mass of a gas if its density at 16°C and 0.980 atm is found to be 2.66 g/L?

12. Calculate the density of a gas if at 30°C it has a pressure of 0.398 atm. Molar mass of the gas is 80 g/mole.

13. How many moles of a gas are present if, when the gas is collected over water at 24°C, it has a total pressure of 755 torr and a volume of 3.5 L? The vapor pressure of water at this temperature is 25 torr.

SOLUTIONS

1. How many grams of sodium hydroxide are needed to prepare 235 mL of a 5.0% (m/v)* solution?

2. How many grams of silver nitrate are contained in 500 mL of a 12.5% (m/v) solution?

3. To what volume do you have to dilute 34.6 grams of calcium chloride to obtain an 8.2% (m/v) solution?

4. What is the percent concentration of the solution obtained when 15.2 grams of nickel(II) nitrate are diluted to 1.5 liters?

5. How many grams of sodium nitrate are needed to prepare 1.45 L of a 0.386 M solution?

6. How many grams of copper(II) iodide are contained in 368 mL of a 0.276 M solution?

7. What is the molarity of the solution obtained when 26.3 g of potassium bromide are diluted to 487 mL?

8. How many grams of lead(II) nitrate are needed to prepare 750. mL of 0.785 M solution?

9. What is the concentration of the solution obtained when 12.5 mL of 0.489 M solution is diluted to 58.5 mL?

10. How many milliliters of 4.5 M solution are needed to prepare 1.46 L of 0.541 M solution?

11. What was the concentration of the original solution if, on diluting 34.6 mL of it to 256 mL, the final concentration was found to be 0.468 M?

12. Which of the following solutions contains the largest number of grams of solute?

    a. 125 mL of 13.2% calcium chloride

    b. 17.4 mL of 1.23 M sodium bromide

    c. 46.7 mL of 0.148 M iron(III) sulfate

---

*(m/v) means mixed percent, grams per 100 mL.

OXIDATION REDUCTION

1. Define:

   a. a redox reaction

   b. a reducing agent

   c. oxidation

2. Determine the oxidation number of the underlined atom in the following species:

   a. H$\underline{C}$O$_3^-$

   b. $\underline{S}$O$_4^{2-}$

   c. $\underline{Mn}$O$_4^-$

   d. H$\underline{N}$O$_2$

   e. $\underline{C}$H$_3$OH

   f. $\underline{P}$O$_3^{3-}$

   g. $\underline{Cr}$O$_5$

   h. H$_2\underline{C}$O

   i. H$_2\underline{S}$

   j. $\underline{Cl}$O$^-$

3. Fill in the electrons (on the proper side!) for the following half-reactions and state whether the reaction is an oxidation or a reduction.

   a. ClO$_2^-$ → ClO$_3^-$

   b. S → SO$_2$

   c. NH$_3$ → NO

   d. BrO$_3$ → BrO$^-$

   e. HCHO → HCOOH

   f. N → NO$_3^-$

   g. PO$_4^{3-}$ → P

   h. AsO$_4^{3-}$ → AsO$_3^{3-}$

ACIDS AND BASES

1. Give the conjugate acid or base of the following species:

Acid	Base
H_2CO_3	
H_2S	
H_2O	
$H_2PO_4^-$	
	CO_3^{2-}
	OH^-
	HSO_4^-
	H_2O

2. Calculate the pH if
   a. $[H^+] = 1.0 \times 10^{-4}$ M
   b. $[OH^-] = 1.0 \times 10^{-8}$ M
   c. $pOH = 9.00$
   d. $[H^+] = 2.3 \times 10^{-6}$ M

3. Calculate the pOH if
   a. $[H^+] = 1.0 \times 10^{-4}$ M
   b. $[OH^-] = 1.0 \times 10^{-3}$ M
   c. $pH = 2.00$
   d. $[OH] = 6.5 \times 10^{-5}$ M

4. Is the solution acidic, basic, or neutral under the following conditions?
   a. $[H^+] = 1.0 \times 10^{-5}$ M
   b. $[H^+] = 1.0 \times 10^{-8}$ M
   c. $[OH^-] = 1.0 \times 10^{-4}$ M
   d. $pH = 3.00$
   e. $pH = 11.00$
   f. $pOH = 7.00$

5. Which of the following solutions is most acidic?
   a. $[H^+] = 1.0 \times 10^{-6}$ M
   b. $[OH] = 1.0 \times 10^{-5}$ M
   c. $pOH = 10.00$
   d. $pH = 3.00$

6. Which of the following solutions is most basic?
   a. $[OH^-] = 1.0 \times 10^{-4}$ M
   b. $pOH = 11.00$
   c. $pH = 5.00$
   d. $[H^+] = 1.0 \times 10^{-12}$ M

Appendix

The Oxidation Numbers of Some Common Cations

Ionic Charge: +1	Ionic Charge: +2	Ionic Charge: +3
Alkali Metals: *Group IA*	*Alkaline Earths:* *Group IIA*	*Group IIIA*
Li^+ — Lithium Na^+ — Sodium K^+ — Potassium Rb^+ — Rubidium Cs^+ — Cesium	Be^{2+} — Beryllium Mg^{2+} — Magnesium Ca^{2+} — Calcium Sr^{2+} — Strontium Ba^{2+} — Barium	Al^{3+} — Aluminum Bi^{3+} — Bismuth Sb^{3+} — Antimony
Transition Elements	*Transition Elements*	*Transition Elements*
Cu^+ — Copper(I) Ag^+ — Silver	Cr^{2+} — Chromium(II) Mn^{2+} — Manganese(II) Fe^{2+} — Iron(II) Co^{2+} — Cobalt(II) Ni^{2+} — Nickel Cu^{2+} — Copper(II) Zn^{2+} — Zinc Cd^{2+} — Cadmium Hg_2^{2+} — Mercury(I) Hg^{2+} — Mercury(II)	Cr^{3+} — Chromium(III) Fe^{3+} — Iron(III) Co^{3+} — Cobalt(III)
Polyatomic Ions		
NH_4^+ — Ammonium		
Others	*Others*	
H^+ — Hydrogen or H_3O^+ — Hydronium	Sn^{2+} — Tin(II) Pb^{2+} — Lead(II)	

The Oxidation Numbers of Some Common Anions

Ionic Charge: –1		Ionic Charge: –2	Ionic Charge: –3
Halogens: *Group VIIA* F^- Fluoride Cl^- Chloride Br^- Bromide I^- Iodide	*Oxyanions* ClO_4^- Perchlorate ClO_3^- Chlorate ClO_2^- Chlorite ClO^- Hypochlorite	*Group VIA* O^{2-} Oxide S^{2-} Sulfide	*Group VA* N^{3-} Nitride P^{3-} Phosphide
Acidic Anions HCO_3^- Hydrogen carbonate HS^- Hydrogen sulfide HSO_4^- Hydrogen sulfate HSO_3^- Hydrogen sulfite $H_2PO_4^-$ Dihydrogen phosphate	BrO_3^- Bromate BrO_2^- Bromite BrO^- Hypobromite IO_4^- Periodate IO_3^- Iodate IO^- Hypoiodite NO_3^- Nitrate NO_2^- Nitrite	*Oxyanions* CO_3^{2-} Carbonate SO_4^{2-} Sulfate SO_3^{2-} Sulfite $C_2O_4^{2-}$ Oxalate CrO_4^{2-} Chromate $Cr_2O_7^{2-}$ Dichromate *Acidic Anion* HPO_4^{2-} Monohydrogen phosphate	*Oxyanion* PO_4^{3-} Phosphate PO_3^{3-} Phosphite
Other Anions SCN^- Thiocyanate CN^- Cyanide H^- Hydride	OH^- Hydroxide $C_2H_3O_2^-$ Acetate MnO_4^- Permanganate	*Diatomic* O_2^{2-} Peroxide	